# A New Era of Thought



Charles Howard Hinton

# INTRODUCTORY NOTE TO PART I.

At the completion of a work, or at the completion of the first part of a work, the feelings are necessarily very different from those with which the work was begun; and the meaning and value of the work itself bear a very different appearance. It will therefore be the simplest and shortest plan, if I tell the reader briefly what the work is to which these pages are a guide, and what I consider to be its value when done.

The task was to obtain a sense of the properties of higher space, or space of four dimensions, in the same way as that by which we reach a sense of our ordinary three-dimensional space. I now prefer to call the task that of obtaining a familiarity with higher matter, which shall be as intuitive to the mind as that of ordinary matter has become. The expression "higher matter" is preferable to "higher space," because it is a somewhat hasty proceeding to split this concrete matter, which we touch and feel, into the abstractions of extension and impenetrability. It seems to me that I cannot think of space without matter, and therefore, as no necessity compels me to such a course, I do not split up the concrete object into subtleties, but I simply ask: "What is that which is to a cube or block or shape of any kind as the cube is to a square?"

In entering upon this inquiry we find the task is twofold. Firstly, there is the theoretical part, which is easy, viz. to set clearly before us the relative conditions which would obtain if there were a matter physically higher than this matter of ours, and to choose the best means of liberating our minds from the limitations imposed on it by the particular conditions under which we are placed. The second part of the task is somewhat laborious, and consists of a constant presentation to the senses of those appearances which portions of higher matter would present, and of a continual dwelling on them, until the higher matter becomes familiar.

The reader must undertake this task, if he accepts it at all, as an experiment. Those of us who have done it, are satisfied that there is that in the results of the experiment which make it well worthy of a trial.

And in a few words I may state the general bearings of this work, for every branch of work has its general bearings. It is an attempt, in the most elementary and simple domain, to pass from the lower to the higher. In pursuing it the mind passes from one kind of intuition to a higher one, and with that transition the horizon of thought is altered. It becomes clear that there is a physical existence transcending the ordinary physical existence; and one becomes inclined to think that the right direction to look is, not away from matter to spiritual existences, but towards the discovery of conceptions of higher matter, and thereby of those material existences whose definite relations to us are apprehended as spiritual intuitions. Thus, "material" would simply mean "grasped by the intellect, become known and familiar." Our apprehension of anything which is not expressed in terms of matter, is vague and indefinite. To realize and live with that which we vaguely discern, we need to apply the intuition of higher matter to the world around us. And this seems to me the great inducement to this study. Let us form our intuition of higher space, and then look out upon the world.

Secondly, in this progress from ordinary to higher matter, as a general type of progress from lower to higher, we make the following observations. Firstly, we become aware that there are certain limitations affecting our regard. Secondly, we discover by our reason what those limitations are, and then force ourselves to go through the experience which would be ours if the limitations did not affect us. Thirdly, we become aware of a capacity within us for transcending those limitations, and for living in the higher mode as we had lived in the previous one.

We may remark that this progress from the ordinary to the higher kind of matter demands an absolute attention to details. It is only in the retention of details that such progress becomes possible. And as, in this question of matter, an absolute and unconventional examination gives us the indication of a higher, so, doubtless, in other questions, if we but come to facts without presupposition, we begin to know that there is a higher and to discover indications of the way whereby we can approach. That way lies in the fulness of detail rather than in the generalization.

Biology has shown us that there is a universal order of forms or organisms, passing from lower to higher. Therein we find an indication that we ourselves take part in this progress. And in using the little cubes we can go through the process ourselves, and learn what it is in a little instance.

But of all the ways in which the confidence gained from this lesson can be applied, the nearest to us lies in the suggestion it gives,—and more than the suggestion, if inclination to think be counted for anything,—in the suggestion of that which is higher than ourselves. We, as individuals, are not the limit and end-all, but there is a higher being than ours. What our relation to it is, we cannot tell, for that is unlike our relation to anything we know. But, perhaps all that happens to us is, could we but grasp it, our relation to it.

At any rate, the discovery of it is the great object beside which all else is as secondary as the routine of mere existence is to companionship. And the method of discovery is full knowledge of each other. Thereby is the higher being to be known. In as much as the least of us knows and is known by another, in so much does he know the higher. Thus, scientific prayer is when two or three meet together, and, in the belief of one higher than themselves, mutually comprehend that vision of the higher, which each one is, and, by absolute fulness of knowledge of the facts of each other's personality, strive to attain a knowledge of that which is to each of their personalities as a higher figure is to its solid sides.

C. H. H.

# PART I.

## INTRODUCTION.

There are no new truths in this book, but it consists of an effort to impress upon and bring home to the mind some of the more modern developments of thought. A few sentences of Kant, a few leading ideas of Gauss and Lobatschewski form the material out of which it is built up.

It may be thought to be unduly long; but it must be remembered that in these times there is a twofold process going on—one of discovery about external nature, one of education, by which our minds are brought into harmony with that which we know. In certain respects we find ourselves brought on by the general current of ideas—we feel that matter is permanent and cannot be annihilated, and it is almost an axiom in our minds that energy is persistent, and all its transformations remains the same in amount. But there are other directions in which there is need of definite training if we are to enter into the thoughts of the time.

And it seems to me that a return to Kant, the creator of modern philosophy, is the first condition. Now of Kant's enormous work only a small part is treated here, but with the difference that should be found between the work of a master and that of a follower. Kant's statements are taken as leading ideas, suggesting a field of work, and it is in detail and manipulation merely that there is an opportunity for workmanship.

Of Kant's work it is only his doctrine of space which is here experimented upon. With Kant the perception of things as being in space is not treated as it seems so obvious to do. We should naturally say that there is space, and there are things in it. From a comparison of those properties which are common to all things we obtain the properties of space. But Kant says that this property of being in space is not so much a quality of any definable objects, as the means by which we obtain an apprehension of definable objects—it is the condition of our mental work.

Now as Kant's doctrine is usually commented on, the negative side is brought into prominence, the positive side is neglected. It is generally said that the mind cannot perceive things in themselves, but can only apprehend them subject to space conditions. And in this way the space conditions are as it were considered somewhat in the light of hindrances, whereby we are prevented from seeing what the objects in themselves truly are. But if we take the statement simply as it is—that we apprehend by means of space—then it is equally allowable to consider our space sense as a positive means by which the mind grasps its experience.

There is in so many books in which the subject is treated a certain air of despondency—as if this space apprehension were a kind of veil which shut us off from nature. But there is no need to adopt this feeling. The first postulate of this book is a full recognition of the fact, that it is by means of space that we apprehend what is. Space is the instrument of the mind.

And here for the purposes of our work we can avoid all metaphysical discussion. Very often a statement which seems to be very deep and abstruse and hard to grasp, is simply the form into which deep thinkers have thrown a very simple and practical observation. And for the present let us look on Kant's great doctrine of space from a practical point of view, and it comes to this—it is important to develop the space sense, for it is the means by which we think about real things.

There is a doctrine which found much favour with the first followers of Kant, that also affords us a simple and practical rule of work. It was considered by Fichte that the whole external world was simply a projection from the *ego*, and the manifold of nature was a recognition by the spirit of itself. What this comes to as a practical rule is, that we can only understand nature in virtue of our own activity; that there is no such thing as mere passive observation, but every act of sight and thought is an activity of our own.

Now according to Kant the space sense, or the intuition of space, is the most fundamental power of the mind. But I do not find anywhere a systematic and thoroughgoing education of the space sense. In every practical pursuit it is needed—in some it is developed. In geometry it is used; but the great reason of failure in education is that, instead of a systematic training of the space sense, it is left to be organized by accident and is called upon to act without having been formed. According to Kant and according to common experience it will be found that a trained thinker is one in whom the space sense has been well developed.

With regard to the education of the space sense, I must ask the indulgence of the reader. It will seem obvious to him that any real pursuit or real observation trains the space sense, and that it is going out of the way to undertake any special discipline.

To this I would answer that, according to my own experience, I was perfectly ignorant of space relations myself before I actually worked at the subject, and that directly I got a true view of space facts a whole series of conceptions, which before I had known merely by repute and grasped by an effort, became perfectly simple and clear to me.

Moreover, to take one instance: in studying the relations of space we always have to do with coloured objects, we always have the sense of weight; for if the things themselves have no weight, there is always a direction of up and down—which implies the sense of weight, and to get rid of these elements requires careful sifting. But perhaps the best point of view to take is this—if the reader has the space sense well developed he will have no difficulty in going through the part of the book which relates to it, and the phraseology will serve him for the considerations which come next.

Amongst the followers of Kant, those who pursued one of the lines of thought in his works have attracted the most attention and have been considered as his successors. Fichte, Schelling, Hegel have developed certain tendencies and have written remarkable books. But the true successors of Kant are Gauss and Lobatchewski.

For if our intuition of space is the means by which we apprehend, then it follows that there may be different kinds of intuitions of space. Who can tell what the absolute space intuition is? This intuition of space must be coloured, so to speak, by the conditions of the being which uses it.

Now, after Kant had laid down his doctrine of space, it was important to investigate how much in our space intuition is due to experience—is a matter of the physical circumstances of the thinking being—and how much is the pure act of the mind.

The only way to investigate this is the practical way, and by a remarkable analysis the great geometers above mentioned have shown that space is not limited as ordinary experience would seem to inform us, but that we are quite capable of conceiving different kinds of space.

Our space as we ordinarily think of it is conceived as limited—not in extent, but in a certain way which can only be realized when we think of our ways of measuring space objects. It is found that there are only three independent directions in which a body can be measured—it must have height, length and breadth, but it has no more than these dimensions. If any other measurement be taken in it, this new measurement will be found to be compounded of the old measurements. It is impossible to find a point in the body which could not be arrived at by travelling in combinations of the three directions already taken.

But why should space be limited to three independent directions?

Geometers have found that there is no reason why bodies should be thus limited. As a matter of fact all the bodies which we can measure are thus limited. So we come to this conclusion, that the space which we use for conceiving ordinary objects in the world is limited to three dimensions. But it might be possible for there to be beings living in a world such that they would conceive a space of four dimensions. All that we can say about such a supposition is, that it is not demanded by our experience. It may be that in the very large or the very minute a fourth dimension of space will have to be postulated to account for parts—but with regard to objects of ordinary magnitudes we are certainly not in a four dimensional world.

And this was the point at which about ten years ago I took up the inquiry.

It is possible to say a great deal about space of higher dimensions than our own, and to work out analytically many problems which suggest themselves. But can we conceive four-dimensional space in the same way in which we can conceive our own space? Can we think of a body in four dimensions as a unit having properties in the same way as we think of a body having a definite shape in the space with which we are familiar?

Now this question, as every other with which I am acquainted, can only be answered by experiment. And I commenced a series of experiments to arrive at a conclusion one way or the other.

It is obvious that this is not a scientific inquiry—but one for the practical teacher.

And just as in experimental researches the skilful manipulator will demonstrate a law of nature, the less skilled manipulator will fail; so here, everything depended on the manipulation. I was not sure that this power lay hidden in the mind, but to put the question fairly would surely demand every resource of the practical art of education.

And so it proved to be; for after many years of work, during which the conception of four-dimensional bodies lay absolutely dark, at length, by a certain change of plan, the whole subject of four-dimensional existence became perfectly clear and easy to impart.

There is really no more difficulty in conceiving four-dimensional shapes, when we go about it the right way, than in conceiving the idea of solid shapes, nor is there any mystery at all about it.

When the faculty is acquired—or rather when it is brought into consciousness, for it exists in every one in imperfect form—a new horizon opens. The mind acquires a development of power, and in this use of ampler space as a mode of thought, a path is opened by using that very truth which, when first stated by Kant, seemed to close the mind within such fast limits. Our perception is subject to the condition of being in space. But space is not limited as we at first think.

The next step after having formed this power of conception in ampler space, is to investigate nature and see what phenomena are to be explained by four-dimensional relations.

But this part of the subject is hardly one for the same worker as the one who investigates how to think in four-dimensional space. The work of building up the power is the work of the practical educator, the work of applying it to nature is the work of the scientific man. And it is not possible to accomplish both tasks at the same time. Consequently the crown is still to be won. Here the method is given of training the mind; it will be an exhilarating moment when an investigator comes upon phenomena which show that external nature cannot be explained except by the assumption of a four-dimension space.

The thought of the past ages has used the conception of a three-dimensional space, and by that means has classified many phenomena and has obtained rules for dealing with matters of great practical utility. The path which opens immediately before us in the future is that of applying the conception of four-dimensional space to the phenomena of nature, and of investigating what can be found out by this new means of apprehension.

In fact, what has been passed through may be called the three-dimensional era; Gauss and Lobatchewski have inaugurated the four-dimensional era.

# CHAPTER I.
# SCEPTICISM AND SCIENCE. BEGINNING OF KNOWLEDGE.

The following pages have for their object to induce the reader to apply himself to the study, in the first place of Space, and then of Higher Space; and, therefore, I have tried by indirect means to show forth those thoughts and conceptions to which the practical work leads.

And I feel that I have a great advantage in this project, inasmuch as many of the thoughts which spring up in the mind of one who studies higher space, and many of the conceptions to which he is driven, turn out to be nothing more nor less than old truths—the property of every mind that thinks and feels—truths which are not generally associated with the scientific apprehension of the world, but which are not for that reason any the less valuable.

And for my own part I cannot do more than put them forward in a very feeble and halting manner. For I have come upon them, not in the way of feeling or direct apprehension, but as the result of a series of works undertaken purely with the desire to know—a desire which did not lift itself to the height of expecting or looking for the beautiful or the good, but which simply asked for something to know.

For I found myself—and many others I find do so also—I found myself in respect to knowledge like a man who is in the midst of plenty and yet who cannot find anything to eat. All around me were the evidences of knowledge—the arts, the sciences, interesting talk, useful inventions—and yet I myself was profited nothing at all; for somehow, amidst all this activity, I was left alone, I could get nothing which I could know.

The dialect was foreign to me—its inner meaning was hidden. If I would, imitating the utterance of my fellows, say a few words, the effort was forced, the whole result was an artificiality, and, if successful, would be but a plausible imposture.

The word "sceptical" has a certain unpleasant association attached to it, for it has been used by so many people who are absolutely certain in a particular line, and attack other people's convictions. But to be sceptical in the real sense is a far more unpleasant state of mind to the sceptic than to any one of his companions. For to a mind that inquires into what it really does know, it is hardly possible to enunciate complete sentences, much less to put before it those complex ideas which have so large a part in true human life.

Every word we use has so wide and fugitive a meaning, and every expression touches or rather grazes fact by so very minute a point, that, if we wish to start with something which we do know, and thence proceed in a certain manner, we are forced away from the study of reality and driven to an artificial system, such as logic or mathematics, which, starting from postulates and axioms, develops a body of ideal truth which rather comes into contact with nature than is nature.

Scientific achievement is reserved for those who are content to absorb into their consciousness, by any means and by whatever way they come, the varied appearances of nature, whence and in which by reflection they find floating as it were on the sea of the unknown, certain similarities, certain resemblances and analogies, by means of which they collect together a body of possible predictions and inferences; and in nature they find correspondences which are actually verified. Hence science exists, although the conceptions in the mind cannot be said to have any real correspondence in nature.

We form a set of conceptions in the mind, and the relations between these conceptions give us relations which we find actually vibrating in the world around us. But the conceptions themselves are essentially artificial.

We have a conception of atoms; but no one supposes that atoms actually exist. We suppose a force varying inversely as the square of the distance; but no one supposes such a mysterious thing to really be in nature. And when we come to the region of descriptive science, when we come to simple observation, we do not find ourselves any better provided with a real knowledge of nature. If, for instance, we think of a plant, we picture to ourselves a certain green shape, of a more or less definite character. This green shape enables us to recognise the plant we think of, and to describe it to a certain extent. But if we inquire into our imagination of it, we find that our mental image very soon diverges from the fact. If, for instance, we cut the plant in half, we find cells and tissues of various kinds. If we examine our idea of the plant, it has merely an external and superficial resemblance to the plant itself. It is a mental drawing meeting the real plant in external appearance; but the two things, the plant and our thought of it, come as it were from different sides—they just touch each other as far as the colour and shape are concerned, but as structures and as living organisms they are as wide apart as possible.

Of course by observation and study the image of a plant which we bear in our minds may be made to resemble a plant as found in the fields more and more. But the agreement with nature lies in the multitude of points superadded on to the notion of greenness which we have at first—there is no natural starting-point where the mind meets nature, and whence they can travel hand in hand.

It almost seems as if, by sympathy and feeling, a human being was easier to know than the simplest object. To know any object, however simple, by the reason and observation requires an endless process of thought and looking, building up the first vague impression into something like in more and more respects. While, on the other hand, in dealing with human beings there is an inward sympathy and capacity for knowing which is independent of, though called into play by, the observation of the actions and outward appearance of the human being.

But for the purpose of knowing we must leave out these human relationships. They are an affair of instinct and inherited unconscious experience. The mind may some day rise to the level of these inherited apprehensions, and be able to explain them; but at present it is far more than overtasked to give an account of the simplest portions of matter, and is quite inadequate to give an account of the nature of a human being.

Asking, then, what there was which I could know, I found no point of beginning. There were plenty of ways of accumulating observations, but none in which one could go hand in hand with nature.

A child is provided in the early part of its life with a provision of food adapted for it. But it seemed that our minds are left without a natural subsistence, for on the one hand there are arid mathematics, and on the other there is observation, and in observation there is, out of the great mass of constructed mental images, but little which the mind can assimilate. To the worker at science of course this crude and omnivorous observation is everything; but if we ask for something which we can know, it is like a vast mass of indigestible material with every here and there a fibre or thread which we can assimilate.

In this perplexity I was reduced to the last condition of mental despair; and in default of finding anything which I could understand in nature, I was sufficiently humbled to learn anything which seemed to afford a capacity of being known.

And the objects which came before me for this endeavour were the simple ones which will be plentifully used in the practical part of this book. For I found that the only assertion I could make about external objects, without bringing in unknown and unintelligible relations, was this: I could say how things were arranged. If a stone lay between two others, that was a definite and intelligible fact, and seemed primary. As a stone itself, it was an unknown somewhat which one could get more and more information about the more one studied the various sciences. But granting that there were some things there which we call stones, the way they were arranged was a simple and obvious fact which could be easily expressed and easily remembered.

And so in despair of being able to obtain any other kind of mental possession in the way of knowledge, I commenced to learn arrangements, and I took as the objects to be arranged certain artificial objects of a simple shape. I built up a block of cubes, and giving each a name I learnt a mass of them.

Now I do not recommend this as a thing to be done. All I can say is that genuinely then and now it seemed and seems to be the only kind of mental possession which one can call knowledge. It is perfectly definite and certain. I could tell where each cube came and how it was related to each of the others. As to the cube itself, I was profoundly ignorant of that; but assuming that as a necessary starting-point, taking that as granted, I had a definite mass of knowledge.

But I do not wish to say that this is better than any kind of knowledge which other people may find come home to them. All I want to do is to take this humble beginning of knowledge and show how inevitably, by devotion to it, it leads to marvellous and far-distant truths, and how, by a strange path, it leads directly into the presence of some of the highest conceptions which great minds have given us.

I do not think it ought to be any objection to an inquiry, that it begins with obvious and common details. In fact I do not think that it is possible to get anything simpler, with less of hypothesis about it, and more obviously a simple taking in of facts than the study of the arrangement of a block of cubes.

Many philosophers have assumed a starting point for their thought. I want the reader to accept a very humble one and see what comes of it. If this leads us to anything, no doubt greater results will come from more ambitious beginnings.

And now I feel that I have candidly exposed myself to the criticism of the reader. If he will have the patience to go on, we will begin and build up on our foundations.

# CHAPTER II.
## APPREHENSION OF NATURE. INTELLIGENCE. STUDY OF ARRANGEMENT OR SHAPE.

Nature is that which is around us. But it is by no means easy to get to nature. The savage living we may say in the bosom of nature, is certainly unapprehensive of it, in fact it has needed the greatness of a Wordsworth and of generations of poets and painters to open our eyes even in a slight measure to the wonder of nature.

Thus it is clear that it is not by mere passivity that we can comprehend nature; it is the goal of an activity, not a free gift.

And there are many ways of apprehending nature. There are the sounds and sights of nature which delight the senses, and the involved harmonies and the secret affinities which poetry makes us feel; then, moreover, there is the definite knowledge of natural facts in which the memory and reason are employed.

Thus we may divide our means of coming into contact with nature into three main channels: the senses, the imagination, and the mind. The imagination is perhaps the highest faculty, but we leave it out of consideration here, and ask: How can we bring our minds into contact with nature?

Now when we see two people of diverse characters we sometimes say that they cannot understand one another—there is nothing in the one by which he can understand the other—he is shut out by a limitation of his own faculties.

And thus our power of understanding nature depends on our own possession; it is in virtue of some mental activity of our own that we can apprehend that outside activity which we call nature. And thus the training to enable us to approach nature with our minds will be some active process on our own part.

In the course of my experience as a teacher I have often been struck by the want of the power of reason displayed by pupils; they are not able to put two and two together, as the saying goes, and I have been at some pains to investigate wherein this curious deficiency lies, and how it can be supplied. And I have found that there is in the curriculum no direct cure for it—the discipline which supplies it is not one which comes into school methods, it is a something which most children obtain in the natural and unsupervised education of their first contact with the world, and lies before any recognised mode of distinction. They can only understand in virtue of an activity of their own, and they have not had sufficient exercise in this activity.

In the present state of education it is impossible to diverge from the ordinary routine. But it is always possible to experiment on children who are out of the common line of education. And I believe I am amply justified by the result of my experiments.

I have seen that the same activity which I have found makes that habit of mind which we call intelligence in a child, is the source of our common and everyday rational intellectual work, and that just as the faculties of a child can be called forth by it, so also the powers of a man are best prepared by the same means, but on an ampler scale.

A more detailed development of the practical work of Part II., would be the best training for the mind of a child. An extension of the work of that Part would be the training which, hand in hand with observation and recapitulation, would best develop a man's thought power.

In order to tell what the activity is by the prosecution of which we can obtain mental contact with nature we should observe what it is which we say we "understand" in any phenomenon of nature which has become clear to us.

When we look at a bright object it seems very different from a dull one. A piece of bright steel hardly looks like the same substance as a piece of dull steel. But the difference of appearance in the two is easily accounted for by the different nature of the surface in the two cases; in the one all the irregularities are done away with, and the rays of light which fall on it are sent off again without being dispersed and broken up. In the case of the dull iron the rays of light are broken up and divided, so that they are not transmitted with intensity in any one direction, but flung off in all sorts of directions.

Here the difference between the bright object and the dull object lies in the arrangement of the particles on its surface and their influence on the rays of light.

Again, with light itself the differences of colour are explained as being the effect on us of rays of different rates of vibration. Now a vibration is essentially this, a series of arrangements of matter which follow each other in a closed order, so that when the set has been run through, the first arrangement follows again. The whole theory of light is an account of arrangements of the particles in the transmitting medium, only the arrangements alter—are not permanent in any one characteristic, but go through a complete cycle of varieties.

Again, when the movements of the heavenly bodies are deduced from the theory of universal gravitation, what we primarily do is to take account of arrangement; for the law of gravity connects the movements which the attracted bodies tend to make with their distances, that is, it shows how their movements depend on their arrangement. And if gravity as a force is to be explained itself, the suppositions which have been put forward resolve it into the effect of the movements of small bodies; that is to say, gravity, if explained at all, is explained as the result of the arrangement and altering arrangements of small particles.

Again, to take the idea which proceeding from Goethe casts such an influence on botanical observation. Goethe (and also Wolf) laid down that the parts of a flower were modified leaves—and traced the stages and intermediate states between the ordinary green leaf and the most gorgeous petal or stamen or carpel, so unlike a leaf in form and function.

Now the essential value in this conception is, that it enables us to look, upon these different organs of a plant as modifications of one and the same organ—it enables us to think about the different varieties of the flower head as modifications of one single plant form. We can trace correspondences between them, and are led to possible explanations of their growth. And all this is done by getting rid of pistil and stamen as separate entities, and looking on them as leaves, and their parts due to different arrangement of the leaf structure. We have reduced these diverse objects to a common element, we have found the unit by whose arrangements the whole is produced. And in this department of thought, as also to take another instance, in chemistry, to understand is practically this: we find units (leaves or atoms) combinations of which account for the results which we see. Thus we see that that which the mind essentially apprehends is arrangement.

And this holds over the whole range of mental work, from the simplest observation to the most complex theory. When the eye takes in the form of an external object there is something more than a sense impression, something more than a sensation of greenness and light and dark. The mind works as well as the sense, and these sense impressions are definitely grouped in what we call the shape of the object. The essential act of perceiving lies in the apprehension of a shape, and a shape is an arrangement of parts. It does not matter what these parts are; if we take meaningless dots of colour and arrange them we obtain a shape which represents the appearance of a stone or a leaf to a certain degree. If we want to make our representation still more like, we must treat each of the dots as in themselves arrangements, we must compose each of them by many strokes and dots of the brush. But even in this case we have not got anything else besides arrangement. The ultimate element, the small items of light and shade or of colour, are in themselves meaningless; it is in their arrangement that the likeness of the representation consists.

Thus, from a drawing to our notion of the planetary system, all our contact with nature lies in this, in an appreciation of arrangement.

Hence to prepare ourselves for the understanding of nature, we must "arrange." In virtue of our activity in making arrangements we prepare ourselves to do what is called understand nature. Or we may say, that which we call understanding nature is to discern something similar in nature to that which we do when we arrange elements into compounded groups.

Now if we study arrangement in the active way, we must have something to arrange; and the things we work with may be either all alike, or each of them varying from every other.

If the elements are not alike then we are not studying pure arrangement; but our knowledge is affected by the compound nature of that with which we deal. If the elements are all alike, we have what we call units. Hence the discipline preparatory for the understanding of nature is the active arrangement of like units.

And this is very much the case with all educational processes; only the things chosen to arrange are in general words, which are so complicated and carry such a train of association that, unless the mind has already acquired a knowledge of arrangement, it is puzzled and hampered, and never gets a clear apprehension of what its work is.

Now what shall we choose for our units? Any unit would do; but it ought to be a real thing—it ought to be something which can be touched and seen, not something which no one has ever touched or seen, and which is even incapable of definition, like a "number."

I would divide studies into two classes: those which create the faculty of arrangement, and those which use it and exercise it. Mathematics exercises it, but I do not think it creates it; and unfortunately, in mathematics as it is now often taught, the pupil is launched , into a vast system of symbols—the whole use and meaning of symbols (namely, as means to acquire a clear grasp of facts) is lost to him.

Of the possible units which will serve, I take the cube; and I have found that whenever I took any other unit I got wrong, puzzled and lost my way. With the cube one does not get along very fast, but everything is perfectly obvious and simple, and builds up into a whole of which every part is evident.

And I must ask the reader to absolutely erase from his mind all desire or wish to be able to predict or assert anything about nature, and he must please look with horror on any mental process by which he gets at a truth in an ingenious but obscure and inexplicable way. Let him take nothing which is not perfectly clear, patent and evident, demonstrable to his senses, a simple repetition of obvious fact.

Our work will then be this: a study, by means of cubes, of the facts of arrangement. And the process of learning will be an active one of actually putting up the cubes. In this way we do for the mind what Wordsworth does for the imagination—we bring it into contact with nature.

# CHAPTER III.
# THE ELEMENTS OF KNOWLEDGE.

There are two elements which enter into our knowledge with respect to any phenomenon.

If, for instance, we take the sun, and ask ourselves what we observe, we notice that it is a bright, moving body; and of these two qualities, the brightness and the movement, each seems equally predicable of the sun. It does move, and it is bright.

Now further study discloses to us that there is a difference between these two affirmations. The motion of the sun in its diurnal course round the earth is only apparent; but it is really a bright, hot body.

Now of these two assertions which the mind naturally makes about the sun, one—that it is moving—depends on the relation of the beholder to the sun, the other is true about the sun itself. The observed motion depends on a fact affecting oneself and having nothing to do with the sun, while the brightness is really a quality of the sun itself.

Now we will call those qualities or appearances which we notice in a body which are due to the particular conditions under which oneself is placed in observing it, the self elements; those facts about it which are independent of the observer's particular relationship we will call the residual element. Thus the sun's motion is a self element in our thought of the sun, its brightness is a residual element.

It is not, of course, possible to draw a line distinctly between the self elements and the residual elements. For instance, some people have denied that brightness is a quality of things, but that it depends on the capacity of the being for receiving sensations; and for brightness they would substitute the assertion that the sun is giving forth a great deal of energy in the form of heat and light.

But there is no object in pursuing the discussion further. The main distinction is sufficiently obvious. And it is important to separate the self elements involved in our knowledge as far as possible, so that the residual elements may be kept for our closer attention. By getting rid of the self elements we put ourselves in a position in which we can propound sensible questions. By getting rid of the notion of its circular motion round the earth we prepare our way to study the sun as it really is. We get the subject clear of complications and extraneous considerations.

It would hardly be worth while to dwell on this consideration were it not of importance in our study of arrangement. But the fact is that directly a subject has been cleared of the self elements, it seems so absurd to have had them introduced at all that the great difficulty there was in getting rid of them is forgotten.

With regard to the knowledge we have at the present day about scientific matters, there do not seem to be any self elements present. But the worst about a self element is, that its presence is never dreamed of till it is got rid of; to know that it is there is to have done away with it. And thus our body of knowledge is like a fluid which keeps clear, not because there are no substances in solution, but because directly they become evident they fall down as precipitates.

Now one of our serious pieces of work will be to get rid of the self elements in the knowledge of arrangement.

And the kind of knowledge which we shall try to obtain will be somewhat different from the kind of knowledge which we have about events or natural phenomena. In the large subjects which generally occupy the mind the things thought of are so complicated that every detail cannot possibly be considered. The principles of the whole are realized, and then at any required time the principles can be worked out. Thus, with regard to a knowledge of the planetary system, it is said to be known if the law of movement of each of the planets is recognized, and their positions at any one time committed to memory. It is not our habit to remember their relative positions with regard to one another at many intervals, so as to have an exhaustive catalogue of them in our minds. But with regard to the elements of knowledge with which we shall work, the subject is so simple that we may justly demand of ourselves that we will know every detail.

And the knowledge we shall acquire will be much more one of the sense and feeling than of the reason. We do not want to have a rule in our minds by which we can recall the positions of the different cubes, but we want to have an immediate apprehension of them. It was Kant who first pointed out how much of thought there was embodied in the sense impressions; and it is this embodied thought which we wish to form.

## CHAPTER IV.
## THEORY AND PRACTICE.

Both in science and in morals there is an important distinction to be drawn between theory and practice. A knowledge of chemistry does not consist in the intellectual appreciation of different theories and principles, but in being able to act in accordance with the facts of chemical combination, so that by means of the appliances of chemistry practical results are produced. And so in morals—the theoretic acquaintance with the principles of human action may consist with a marked degree of ignorance of how to act amongst other human beings.

Now the use of the word "learn" has been much restricted to merely theoretic studies. It requires to be enlarged to the scientific meaning. And to know, should include practice in actual manipulation.

Let us take an instance. We all know what justice is, and any child can be taught to tell the difference between acting justly and acting unjustly. But it is a different thing to teach them to act with justice. Something is done which affects them unpleasantly. They feel an impulse to retaliate. In order to see what justice demands they have to put their personal feeling on one side. They have to get rid of those conditions under which they apprehended the effects of the action at first, and they have to look on it from another point of view. Then they have to act in accordance with this view.

Now there are two steps—one an intellectual one of understanding, one a practical one of carrying out the view. Neither is a moral step. One demands intelligence, the other the formation of a habit, and this habit can be inculcated by precept, by reward and punishment, by various means. But as human nature is constituted, if the habit of justice is inculcated it touches a part of the being. There is an emotional response. We know but little of a human being, but we can safely say that there are depths in it, beyond the feelings of momentary resentment and the stimulus of pleasurable or painful sensation, to which justice is natural.

How little adequate is our physical knowledge of a human being as a bodily frame to explain the fact of human life. Now and again we see one of these isolated beings bound up in another, as if there was an undiscovered physical bond between them. And in all there is this sense of justice—a kind of indwelling verdict of the universal mind, if we may use such an expression, in virtue of which a man feels not as a single individual but as all men.

With respect to justice, it is not only necessary to take the view of one other person than oneself, but that of many. There may be justice which is very good justice from the point of view of a party, but very bad justice from the point of view of a nation. And if we suppose an agency outside the human race, gifted with intelligence, and affecting the race, in the way for instance of causing storms or disturbances of the ground, in order to judge it with justice we should have to take a standpoint outside the race of men altogether. We could not say that this agency was bad. We should have to judge it with reference to its effect on other sentient beings.

There are some words which are often used in contrast with each other—egoism and altruism; and each seems to me unmeaning except as terms in a contrast.

Let us take an instance. A boy has a bag of cakes, and is going to enjoy them by himself. His parent stops him, and makes him set up some stumps and begin to learn to play cricket with another boy. The enjoyment of the cakes is lost—he has given that up; but after a little while he has a pleasure which is greater than that of cakes in solitude. He enters into the life of the game. He has given up, or been forced to give up, the pleasure he knew, and he has found a greater one. What he thought about himself before was that he liked cakes, now what he thinks about himself is that he likes cricket. Which of these is the true thought about himself? Neither, probably, but at any rate it is more near the truth to say that he likes the cricket. If now we use the word self to mean that which a person knows of himself, and it is difficult to see what other meaning it can have, his self as he knew it at first was thwarted, was given up, and through that he discovered his true self. And again with the cricket; he will make the sacrifice of giving that up, voluntarily or involuntarily, and will find a truer self still.

In general there is not much difficulty in making a boy find out that he likes cricket; and it is quite possible for him to eat his cakes first and learn to play cricket afterwards—the cricket will not come to him as a thwarting in any sense of what he likes better. But this ease in entering in to the pursuit only shows that the boy's nature is already developed to the level of enjoying the game. The distinct moral advance would come in such a case when something which at first was hard to him to do was presented to him—and the hardness, the unpleasantness is of a double kind, the giving up of a pursuit or indulgence to which he is accustomed, and the exertion of forming the habits demanded by the new pursuit.

Now it is unimportant whether the renunciation is forced or willingly taken. But as a general rule it may be laid down, that by giving up his own desires as he feels them at the moment, to the needs and advantage of those around him, or to the objects which he finds before him demanding accomplishment, a human being passes to the discovery of his true self on and on. The process is limited by the responsibilities which a man finds come upon him.

The method of moral advance is to acquire a practical knowledge; he must first see what the advantage of some one other than himself would be, and then he must act in accordance with that view of things. Then having acted and formed a habit, he discovers a response in himself. He finds that he really cares, and that his former limited life was not really himself. His body and the needs of his body, so far as he can observe them, externally are the same as before; but he has obtained an inner and unintellectual, but none the less real, apprehension of what he is.

Thus altruism, or the sacrifice of egoism to others, is followed by a truer egoism, or assertion of self, and this process flashed across by the transcendent lights of religion, wherein, as in the sense of justice and duty, untold depths in the nature of man are revealed entirely unexpressed by the intellectual apprehension which we have of him as an animal frame of a very high degree of development, is the normal one by which from childhood a human being develops into the full responsibilities of a man.

Now both in science and in conduct there are self elements. In science, getting rid of the self elements means a truer apprehension of the facts about one; in conduct, getting rid of the self elements means obtaining a truer knowledge of what we are—in the way of feeling more strongly and deeply and being bound and linked in a larger scale.

Thus without pretending to any scientific accuracy in the use of terms, we can assign a certain amount of meaning to the expression—getting rid of self elements. And all that we can do is to take the rough idea of this process, and then taking our special subject matter, apply it. In affairs of life experiments lead to disaster. But happily science is provided wherein the desire to put theories into practice can be safely satisfied—and good results sometimes follow. Were it not for this the human race might before now have been utopiad from off the face of the earth.

In experiment, manipulation is everything; we must be certain of all our conditions, otherwise we fail assuredly and have not even the satisfaction of knowing that our failure is due to the wrongness of our conjectures.

And for our purposes we use a subject matter so simple that the manipulation is easy.

# CHAPTER V.
# KNOWLEDGE: SELF-ELEMENTS.

I must now go with somewhat of detail into the special subject in which these general truths will be exhibited. Everything I have to say would be conceived much more clearly by a very little practical manipulation.

But here I want to put the subject in as general a light as possible, so that there may be no hindrance to the judgment of the reader.

And when I use the word "know," I assume something else than the possession of a rule, by which it can be said how facts are. By knowing I mean that the facts of a subject all lie in the mind ready to come out vividly into consciousness when the attention is directed on them. Michael Angelo knew the human frame, he could tell every little fact about it; if he chose to call up the image, he would see mentally how each muscle and fold of the skin lay with regard to the surrounding parts. We want to obtain a knowledge as good as Michael Angelo's. There is a great difference between Michael Angelo and us; but let that difference be expressed, not in our way of knowing, but in the difference between the things he knew and the things we know. We take a very simple structure and know it as absolutely as he knew the complicated structure of the human body.

And let us take a block of cubes; any number will do, but for convenience sake let us take a set of twenty-seven cubes put together so as to form a large cube of twenty-seven parts. And let each of these cubes be marked so as to be recognized, and let each have a name so that it can be referred to. And let us suppose that we have learnt this block of cubes so that each one is known—that is to say, its position in the block is known and its relation to the other blocks.

Now having obtained this knowledge of the block as it stands in front of us, let us ask ourselves if there is any self element present in our knowledge of it.

And there is obviously this self element present. We have learnt the cubes as they stand in accordance with our own convenience in putting them up. We put the lowest ones first, and the others on the top of them, and we distinctly conceive the lower ones as supporting the upper ones. Now this fact of support has nothing to do with the block of cubes itself, it depends on the conditions under which we come to apprehend the block of cubes, it depends on our position on the surface of the earth, whereby gravity is an all important factor in our experience. In fact our sight has got so accustomed to take gravity into consideration in its view of things, that when we look at a landscape or object with our head upside down we do not see it inverted, but we superinduce on the direct sense impressions our knowledge of the action of gravity, and obtain a view differing very little from what we see when in an upright position.

It will be found that every fact about the cubes has involved in it a reference to up and down. It is by being above or below that we chiefly remember where the cubes are. But above and below is a relation which depends simply on gravity. If it were not for gravity above and below would be interchangeable terms, instead of expressing a difference of marked importance to us under our conditions of existence. Now we put "being above" or "being below" into the cubes themselves and feel it a quality in them—it defines their position. But this above or below really comes from the conditions in which we are. It is a self element, and as such, to obtain a true knowledge of the cubes we must get rid of it.

And now, for the sake of a process which will be explained afterwards, let us suppose that we cannot move the block of cubes which we have put up. Let us keep it fixed.

In order to learn how it is independent of gravity the best way would be to go to a place where gravity has virtually ceased to act; at the centre of the earth, for instance, or in a freely falling shell.

But this is impossible, so we must choose another way. Let us, then, since we cannot get rid of gravity, see what we have done already. We have learnt the cubes, and however they are learnt, it will be found that there is a certain set of them round which the others are mentally grouped, as being on the right or left, above or below. Now to get our knowledge as perfect as we can before getting rid of the self element up and down, we have to take as central cubes in our mind different sets again and again, until there are none which are primary to us.

Then there remains only the distinction of some being above others. Now this can only be made to sink out of the primary place in our thoughts by reversing the relation. If we turned the block upside down, and learnt it in this new position, then we should learn the position of the cubes with regard to each other with that element in them, which comes from the action of gravity, reversed. And the true nature of the arrangement to which we added something in virtue of our sensation of up and down, would become purer and more isolated in our minds.

We have, however, supposed that the cubes are fixed. Then, in order to learn them, we must put up another block showing what they would be like in the supposed new position. We then take a set of cubes, models of the original cubes, and by consideration we can put them in such positions as to be an exact model of what the block of cubes would be if turned upside down.

And here is the whole point on which the process depends. We can tell where each cube would come, but we do not *know* the block in this new position. I draw a distinction between the two acts, "to tell where it would be," and to "know." Telling where it would be is the preparation for knowing. The power of assigning the positions may be called the theory of the block. The actual knowledge is got by carrying out the theory practically, by putting up the blocks and becoming able to realize without effort where each one is.

It is not enough to put up the model blocks in the reverse position. It is found that this up and down is a very obstinate element indeed, and a good deal of work is requisite to get rid of it completely. But when it is got rid of in one set of cubes, the faculty is formed of appreciating shape independently of the particular parts which are above or below on first examination. We discover in our own minds the faculty of appreciating the facts of position independent of gravity and its influence on us. I have found a very great difference in different minds in this respect. To some it is easy, to some it is hard.

And to use our old instance, the discovery of this capacity is like the discovery of a love of justice in the being who has forced himself to act justly. It is a capacity for being able to take a view independent of the conditions under which he is placed, and to feel in accordance with that view. There is, so far as I know, no means of arriving immediately at this impartial appreciation of shape. It can only be done by, as it were, extending our own body so as to include certain cubes, and appreciating then the relation of the other cubes to those. And after this, by identifying ourselves with other cubes, and in turn appreciating the relation of the other cubes to these. And the practical putting up of the cubes is the way in which this power is gained. It springs up with a repetition of the mechanical acts. Thus there are three processes. 1st, An apprehension of what the position of the cubes would be. 2nd, An actual putting of them up in accordance with that apprehension, 3rd, The springing up in the mind of a direct feeling of what the block is, independent of any particular presentation.

Thus the self element of up and down can be got rid of out of a block of cubes.

And when even a little block is known like this, the mind has gained a great deal.

Yet in the apprehension and knowledge of the block of cubes with the up and down relation in them, there is more than in the absolute apprehension of them. For there is the apprehension of their position and also of the effect of gravity on them in their position.

Imagine ourselves to be translated suddenly to another part of the universe, and to find there intelligent beings, and to hold conversation with them. If we told them that we came from a world, and were to describe the sun to them, saying that it was a bright, hot body which moved round us, they would reply: You have told us something about the sun, but you have also told us something about yourselves.

Thus in the apprehension of the sun as a body moving round us there is more than in the apprehension of it as not moving round, for we really in this case apprehend two things—the sun and our own conditions. But for the purpose of further knowledge it is most important that the more abstract knowledge should be acquired. The self element introduced by the motion of the earth must be got rid of before the true relations of the solar system can be made out.

And in our block of cubes, it will be found that feelings about arrangement, and knowledge of space, which are quite unattainable with our ordinary view of position, become simple and clear when this discipline has been gone through.

And there can be no possible mental harm in going through this bit of training, for all that it comes to is looking at a real thing as it actually is—turning it round and over and learning it from every point of view.

# CHAPTER VI.
# FUNCTION OF MIND. SPACE AGAINST METAPHYSICS. SELF-LIMITATION AND ITS TEST. A PLANE WORLD.

We now pass on to the question: Are there any other self elements present in our knowledge of the block of cubes?

When we have learnt to free it from up and down, is there anything else to be got rid of?

It seems as if, when the cubes were thus learnt, we had got as abstract and impersonal a bit of knowledge as possible.

But, in reality, in the relations of the cubes as we thus apprehend them there is present a self element to which the up and down is a mere trifle. If we think we have got absolute knowledge we are indeed walking on a thin crust in unconsciousness of the depths below.

We are so certain of that which we are habituated to, we are so sure that the world is made up of the mechanical forces and principles which we familiarly deal with, that it is more of a shock than a welcome surprise to us to find how mistaken we were.

And after all, do we suppose that the facts of distance and size and shape are the ultimate facts of the world—is it in truth made up like a machine out of mechanical parts? If so, where is there room for that other which we know—more certainly, because inwardly—that reverence and love which make life worth having? No; these mechanical relations are our means of knowing about the world; they are not reality itself, and their primary place in our imaginations is due to the familiarity which we have with them, and to the peculiar limitations under which we are.

But I do not for a moment wish to go in thought beyond physical nature—I do not suppose that in thought we can. To the mind it is only the body that appears, and all that I hope to do is to show material relations, mechanism, arrangements.

But much depends on what kind of material relations we perceive outside us. A human being, an animal and a machine are to the mind all merely portions of matter arranged in certain ways. But the mind can give an exhaustive account of the machine, account fairly well for the animal, while the human being it only defines externally, leaving the real knowledge to be supplied by other faculties.

But we must not under-estimate the work of the mind, for it is only by the observation of and thought about the bodies with which we come into contact that we know human beings. It is the faculty of thought that puts us in a position to recognize a soul.

And so, too, about the universe—it is only by correct thought about it that we can perceive its true moral nature.

And it will be found that the deadness which we ascribe to the external world is not really there, but is put in by us because of our own limitations. It is really the self elements in our knowledge which make us talk of mechanical necessity, dead matter. When our limitations fall, we behold the spirit of the world like we behold the spirit of a friend—something which is discerned in and through the material presentation of a body to us.

Our thought means are sufficient at present to show us human souls; but all except human beings is, as far as science is concerned, inanimate. One self element must be got rid of from our perception, and this will be changed.

The one thing necessary is, that in matters of thinking we will not admit anything that is not perfectly clear, palpable and evident. On the mind the only conceivable demand is to seek for facts. The rock on which so many systems of philosophy have come to grief is the attempt to put moral principles into nature. Our only duty is to accept what we find. Man is no more the centre of the moral world than he is of the physical world. Then relegate the intellect to its right position of dealing with facts of arrangement—it can appreciate structure—and let it simply look on the world and report on it. We have to choose between metaphysics and space thought. In metaphysics we find lofty ideals—principles enthroned high in our souls, but which reduce the world to a phantom, and ourselves to the lofty spectators of an arid solitude. On the other hand, if we follow Kant's advice, we use our means and find realities linked together, and in the physical interplay of forces and connexion of structure we behold the relations between spirits—those dwelling in man and those above him.

It is difficult to explain this next self element that has to be removed from the block of cubes; it requires a little careful preparation, in fact our language hardly affords us the means. But it is possible to approach indirectly, and to detect the self-element by means of an analogy.

If we suspect there be some condition affecting ourselves which make us perceive things not as they are, but falsely, then it is possible to test the matter by making the supposition of other beings subject to certain conditions, and then examining what the effect on their experience would be of these conditions.

Thus if we make up the appearances which would present themselves to a being subject to a limitation or condition, we shall find that this limitation or condition, when unrecognized by him, presents itself as a general law of his outward world, or as properties and qualities of the objects external to him. He will, moreover, find certain operations possible, others impossible, and the boundary line between the possible and impossible will depend quite as much on the conditions under which he is as on the nature of the operations.

And if we find that in our experience of the outward world there are analogous properties and qualities of matter, analogous possibilities and impossibilities, then it will show to us that we in our turn are under analogous limitations, and that what we perceive as the external world is both the external world and our own conditions. And the task before us will be to separate the two. Now the problem we take up here is this—to separate the self elements from the true fact. To separate them not merely as an outward theory and intelligent apprehension, but to separate them in the consciousness itself, so that our power of perception is raised to a higher level. We find out that we are under limitations. Our next step is to so familiarize ourselves with the real aspect of things, that we perceive like beings not under our limitations. Or more truly, we find that inward soul which itself not subject to these limitations, is awakened to its own natural action, when the verdicts conveyed to it through the senses are purged of the self elements introduced by the senses.

Everything depends on this—Is there a native and spontaneous power of apprehension, which springs into activity when we take the trouble to present to it a view from which the self elements are eliminated? About this every one must judge for himself. But the process whereby this inner vision is called on is a definite one.

And just as a human being placed in natural human relationships finds in himself a spontaneous motive towards the fulfilment of them, discovers in himself a being whose motives transcend the limits of bodily self-regard, so we should expect to find in our minds a power which is ready to apprehend a more absolute order of fact than that which comes through the senses.

I do not mean a theoretical power. A theory is always about it, and about it only. I mean an inner view, a vision whereby the seeing mind as it were identifies itself with the thing seen. Not the tree of knowledge, but of the inner and vital sap which builds up the tree of knowledge.

And if this point is settled, it will be of some use in answering the question: What are we? Are we then bodies only? This question has been answered in the negative by our instincts. Why should we despair of a rational answer? Let us adopt our space thought and develop it.

The supposition which we must make is the following. Let us imagine a smooth surface—like the surface of a table; but let the solid body at which we are looking be very thin, so that our surface is more like the surface of a thin sheet of metal than the top of a table.

And let us imagine small particles, like particles of dust, to lie on this surface, and to be attracted downwards so that they keep on the surface.

But let us suppose them to move freely over the surface. Let them never in their movements rise one over the other; let them all singly and collectively be close to the surface. And let us suppose all sorts of attractions and repulsions between these atoms, and let them have all kinds of movements like the atoms of our matter have.

Then there may be conceived a whole world, and various kinds of beings as formed out of this matter. The peculiarity about this world and these beings would be, that neither the inanimate nor the animate members of it would move away from the surface. Their movements would all lie in one plane, a plane parallel to and very near the surface on which they are.

And if we suppose a vast mass to be formed out of these atoms, and to lie like a great round disk on the surface, compact and cohering closely together, then this great disk would afford a support for the smaller shapes, which we may suppose to be animate beings. The smaller shapes would be attracted to the great disk, but would be arrested at its rim. They would tend to the centre of the disk, but be unable to get nearer to the centre than its rim.

Thus, as we are attracted to the centre of the earth, but walk on its surface, the beings on this disk would be attracted to its centre, but walk on its rim. The force of attraction which they would feel would be the attraction of the disk. The other force of attraction, acting perpendicularly to the plane which keeps them and all the matter of their world to the surface, they would know nothing about. For they cannot move either towards this force or away from it; and the surface is quite smooth, so that they feel no friction in their movement over it.

Now let us realize clearly one of these beings as he proceeds along the rim of his world. Let us imagine him in the form of an outline of a human being, with no thickness except that of the atoms of his world. As to the mode in which he walks, we must imagine that he proceeds by springs or hops, because there would be no room for his limbs to pass each other.

Imagine a large disk on the table before you, and a being, such as the one described, proceeding round it. Let there be small movable particles surrounding him, which move out of his way as he goes along, and let these serve him for respiration; let them constitute an atmosphere.

Forwards and backwards would be to such a being direction along the rim—the direction in which he was proceeding and its reverse.

Then up and down would evidently be the direction away from the disk's centre and towards it. Thus backwards and forwards, up and down, would both lie in the plane in which he was.

And he would have no other liberty of movement except these. Thus the words right and left would have no meaning to him. All the directions in which he could move, or could conceive movement possible, would be exhausted when he had thought of the directions along the rim and at right angles to it, both in the plane.

What he would call solid bodies, would be groups of the atoms of his world cohering together. Such a mass of atoms would, we know, have a slight thickness; namely, the thickness of a single atom. But of this he would know nothing. He would say, "A solid body has two dimensions—height (by how much it goes away from the rim) and thickness (by how much it lies along the rim)." Thus a solid would be a two-dimensional body, and a solid would be bounded by lines. Lines would be all that he could see of a solid body.

Thus one of the results of the limitations under which he exists would be, that he would say, "There are only two dimensions in real things."

In order for his world to be permanent, we must suppose the surface on which he is to be very compact, compared to the particles of his matter; to be very rigid; and, if he is not to observe it by the friction of matter moving on it, to be very smooth. And if it is very compact with regard to his matter, the vibrations of the surface must have the effect of disturbing the portions of his matter, and of separating compound bodies up into simpler ones.

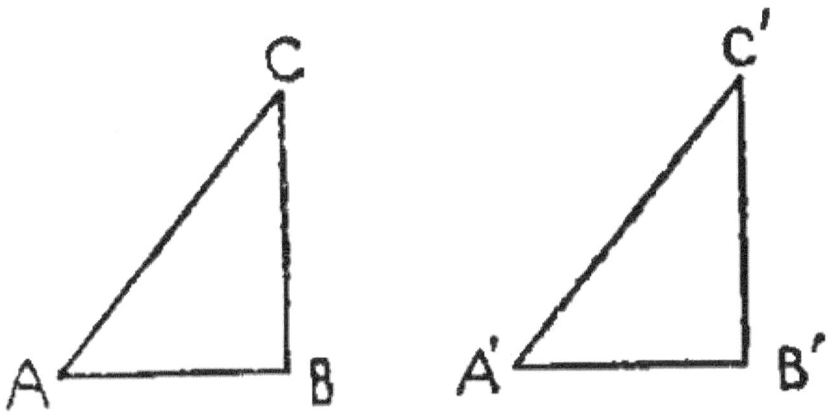


Fig. 1.

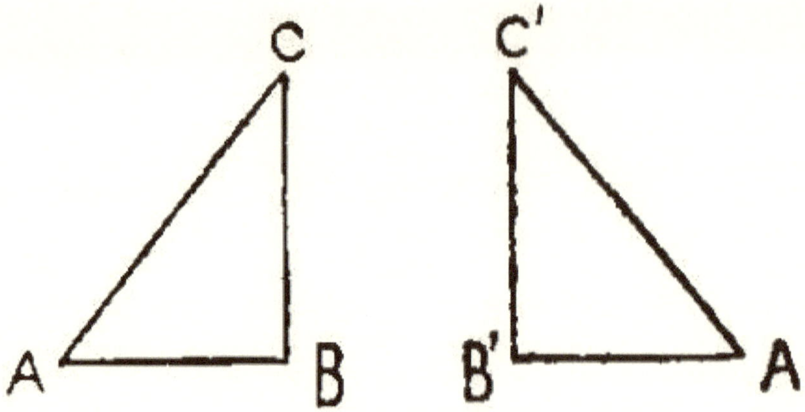


Fig. 2.

Another consequence of the limitation under which this being lies, would be the following:—If we cut out from the corners of a piece of paper two triangles, A B C and A' B' C', and suppose them to be reduced to such a thinness that they are capable of being put on to the imaginary surface, and of being observed by the flat being like other bodies known to him; he will, after studying the bounding lines, which are all that he can see or touch, come to the conclusion that they are equal and similar in every respect; and he can conceive the one occupying the same space as the other occupies, without its being altered in any way.

If, however, instead of putting down these triangles into the surface on which the supposed being lives, as shown in Fig. 1, we first of all turn one of them over, and then put them down, then the plane-being has presented to him two triangles, as shown in Fig. 2.

And if he studies these, he finds that they are equal in size and similar in every respect. But he cannot make the one occupy the same space as the other one; this will become evident if the triangles be moved about on the surface of a table. One will not lie on the same portion of the table that the other has marked out by lying on it.

Hence the plane-being by no means could make the one triangle in this case coincide with the space occupied by the other, nor would he be able to conceive the one as coincident with the other.

The reason of this impossibility is, not that the one cannot be made to coincide, but that before having been put down on his plane it has been turned round. It has been turned, using a direction of motion which the plane-being has never had any experience of, and which therefore he cannot use in his mental work any more than in his practical endeavours.

Thus, owing to his limitations, there is a certain line of possibility which he cannot overstep. But this line does not correspond to what is actually possible and impossible. It corresponds to a certain condition affecting him, not affecting the triangle. His saying that it is impossible to make the two triangles coincide, is an assertion, not about the triangles, but about himself.

Now, to return to our own world, no doubt there are many assertions which we make about the external world which are really assertions about ourselves. And we have a set of statements which are precisely similar to those which the plane-being would make about his surroundings.

Thus, he would say, there are only two independent directions; we say there are only three.

He would say that solids are bounded by lines; we say that solids are bounded by planes.

Moreover, there are figures about which we assert exactly the same kind of impossibility as his plane-being did about the triangles in Fig. 2.

We know certain shapes which are equal the one to the other, which are exactly similar, and yet which we cannot make fit into the same portion of space, either practically or by imagination.

If we look at our two hands we see this clearly, though the two hands are a complicated case of a very common fact of shape. Now, there is one way in which the right hand and the left hand may practically be brought into likeness. If we take the right-hand glove and the left-hand glove, they will not fit any more than the right hand will coincide with the left hand. But if we turn one glove inside out, then it will fit. Now, to suppose the same thing done with the solid hand as is done with the glove when it is turned inside out, we must suppose it, so to speak, pulled through itself. If the hand were inside the glove all the time the glove was being turned inside out, then, if such an operation were possible, the

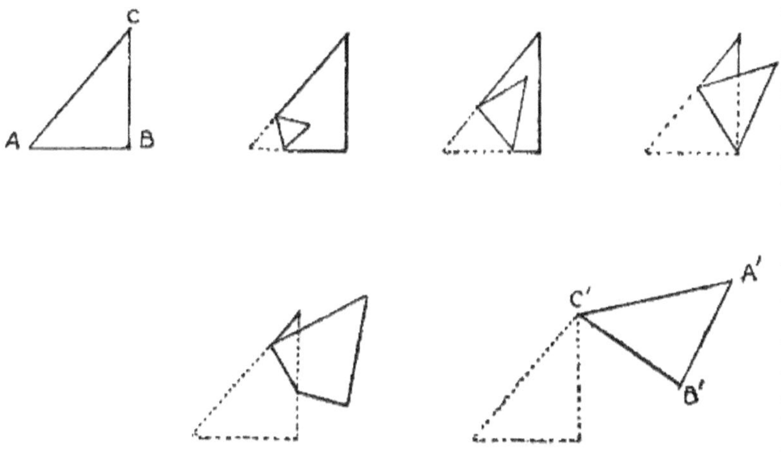

right hand would be turned into an exact model of the left hand. Such an operation is impossible. But curiously enough there is a precisely similar operation which, if it were possible, would, in a plane, turn the one triangle in Fig. 2 into the exact copy of the other.

Look at the triangle in Fig. 2, A B C, and imagine the point A to move into the interior of the triangle and to pass through it, carrying after it the parts of the lines A B and A C to which it is attached, we should have finally a triangle A B C, which was quite like the other of the two triangles A' B' C' in Fig. 2.

Thus we know the operation which produces the result of the "pulling through" is not an impossible one when the plane-being is concerned. Then may it not be that there is a way in which the results of the impossible operation of pulling a hand through could be performed? The question is an open one. Our feeling of it being impossible to produce this result in any way, may be because it really is impossible, or it may be a useful bit of information about ourselves.

Now at this point my special work comes in. If there be really a four-dimensional world, and we are limited to a space or three-dimensional view, then either we are absolutely three-dimensional with no experience at all or capacity of apprehending four-dimensional facts, or we may be, as far as our outward experience goes, so limited; but we may really be four-dimensional beings whose consciousness is by certain undetermined conditions limited to a section of the real space.

Thus we may really be like the plane-beings mentioned above, or we may be in such a condition that our perceptions, not ourselves, are so limited. The question is one which calls for experiment.

We know that if we take an animal, such as a dog or cat, we can by careful training, and by using rewards and punishment, make them act in a certain way, in certain defined cases, in accordance with justice; we can produce the mechanical action. But the feeling of justice will not be aroused; it will be but a mere outward conformity. But a human being, if so trained, and seeing others so acting, gets a feeling of justice.

Now, if we are really four-dimensional, by going through those acts which correspond to a four-dimensional experience (so far as we can), we shall obtain an apprehension of four-dimensional existence—not with the outward eye, but essentially with the mind.

And after a number of years of experiment which were entirely nugatory, I can now lay it down as a verifiable fact, that by taking the proper steps we can feel four-dimensional existence, that the human being somehow, and in some way, is not simply a three-dimensional being—in what way it is the province of science to discover. All that I shall do here is, to put forward certain suppositions which, in an arbitrary and forced manner, give an outline of the relation of our body to four-dimensional existence, and show how in our minds we have faculties by which we recognise it.

## CHAPTER VII.
## SELF ELEMENTS IN OUR CONSCIOUSNESS.

It is often taken for granted that our consciousness of ourselves and of our own feelings has a sort of direct and absolute value.

It is supposed to afford a testimony which does not require to be sifted like our consciousness of external events. But in reality it needs far more criticism to be applied to it than any other mode of apprehension.

To a certain degree we can sift our experience of the external world, and divide it into two portions. We can determine the self elements and the realities. But with regard to our own nature and emotions, the discovery which makes a science possible has yet to be made.

There are certain indications, however, springing from our observation of our own bodies, which have a certain degree of interest.

It is found that the processes of thought and feeling are connected with the brain. If the brain is disturbed, thoughts, sights, and sounds come into the consciousness which have no objective cause in the external world. Hence we may conclusively say that the human being, whatever he is, is in contact with the brain, and through the brain with the body, and through the body with the external world.

It is the structures and movements in the brain which the human being perceives. It is by a structure in the brain that he apprehends nature, not immediately. The most beautiful sights and sounds have no effect on a human being unless there is the faculty in the brain of taking them in and handing them on to the consciousness.

Hence, clearly, it is the movements and structure of the minute portions of matter forming the brain which the consciousness perceives. And it is only by models and representations made in the stuff of the brain that the mind knows external changes.

Now, our brains are well furnished with models and representations of the facts and events of the external world.

But a most important fact still requires its due weight to be laid upon it.

These models and representations are made on a very minute scale—the particles of brain matter which form images and representations are beyond the power of the microscope in their minuteness. Hence the consciousness primarily apprehends the movements of matter of a degree of smallness which is beyond the power of observation in any other way.

Hence we have a means of observing the movements of the minute portions of matter. Let us call those portions of the brain matter which are directly instrumental in making representations of the external world—let us call them brain molecules.

Now, these brain molecules are very minute portions of matter indeed; generally they are made to go through movements and form structures in such a way as to represent the movements and structures of the external world of masses around us.

But it does not follow that the structures and movements which they perform of their own nature are identical with the movements of the portions of matter which we see around us in the world of matter.

It may be that these brain molecules have the power of four-dimensional movement, and that they can go through four-dimensional movements and form four-dimensional structures.

If so, there is a practical way of learning the movements of the very small particles of matter—by observing, not what we can see, but what we can think.

For, suppose these small molecules of the brain were to build up structures and go through movements not in accordance with the rule of representing what goes on in the external world, but in accordance with their own activity, then they might go through four-dimensional movements and form four-dimensional structures.

And these movements and structures would be apprehended by the consciousness along with the other movements and structures, and would seem as real as the others—but would have no correspondence in the external world.

They would be thoughts and imaginations, not observations of external facts.

Now, this field of investigation is one which requires to be worked at.

At present it is only those structures and movements of the brain molecules which correspond to the realities of our three-dimensional space which are in general worked at consistently. But in the practical part of this book it will be found that by proper stimulus the brain molecules will arrange themselves in structures representing a four-dimensional existence. It only requires a certain amount of care to build up mental models of higher space existences. In fact, it is probably part of the difficulty of forming three-dimensional brain models, that the brain molecules have to be limited in their own freedom of motion to the requirements of the limited space in which our practical daily life is carried on.

*Note.*—For my own part I should say that all those confusions in remembering which come from an image taking the place of the original mental model—as, for instance, the difficulty in remembering which way to turn a screw, and the numerous cases of images in thought transference—may be due to a toppling over in the brain, four-dimensionalwise, of the structures formed—which structures would be absolutely safe from being turned into image structures if the brain molecules moved only three-dimensionalwise.

It is remarkable how in science "explaining" means the reference of the movements and tendencies to movement of the masses about us to the movements and tendencies to movement of the minute portions of matter.

Thus, the behaviour of gaseous bodies—the pressure which they exert, the laws of their cooling and intermixture are explained by tracing the movements of the very minute particles of which they are composed.

# CHAPTER VIII.
# RELATION OF LOWER TO HIGHER SPACE. THEORY OF THE ÆTHER.

At this point of our inquiries the best plan is to turn to the practical work, and try if the faculty of thinking in higher space can be awakened in the mind.

The general outline of the method is the same as that which has been described for getting rid of the limitation of up and down from a block of cubes. We supposed that the block was fixed; and to get the sense of what it would be when gravity acted in a different way with regard to it, we made a model of it as it would be under the new circumstances. We thought out the relations which would exist; and by practising this new arrangement we gradually formed the direct apprehension.

And so with higher-space arrangements. We cannot put them up actually, but we can say how they would look and be to the touch from various sides. And we can put up the actual appearances of them, not altogether, but as models succeeding one another; and by contemplation and active arrangement of these different views we call upon our inward power to manifest itself.

In preparing our general plan of work, it is necessary to make definite assumptions with regard to our world, our universe, or we may call it our space, in relation to the wider universe of four-dimensional space.

What our relation to it may be, is altogether undetermined. The real relationship will require a great deal of study to apprehend, and when apprehended will seem as natural to us as the position of the earth among the other planets does to us now.

But we have not got to wait for this exploration in order to commence our work of higher-space thought, for we know definitely that whatever our real physical relationship to this wider universe may be, we are practically in exactly the same relationship to it as the creature we have supposed living on the surface of a smooth sheet is to the world of threefold space.

And this relationship of a surface to a solid or of a solid, as we conjecture, to a higher solid, is one which we often find in nature. A surface is nothing more nor less than the relation between two things. Two bodies touch each other. The surface is the relationship of one to the other.

Again, we see the surface of water.

Thus our solid existence may be the contact of two four-dimensional existences with each other; and just as sensation of touch is limited to the surface of the body, so sensation on a larger scale may be limited to this solid surface.

And it is a fact worthy of notice, that in the surface of a fluid different laws obtain from those which hold throughout the mass. There are a whole series of facts which are grouped together under the name of surface tensions, which are of great importance in physics, and by which the behaviour of the surfaces of liquids is governed.

And it may well be that the laws of our universe are the surface tensions of a higher universe.

But these expressions, it is evident, afford us no practical basis for investigation. We must assume something more definite, and because more definite (in the absence of details drawn from experience), more arbitrary.

And we will assume that the conditions under which we human beings are, exactly resemble those under which the plane-beings are placed, which have been described.

This forms the basis of our work; and the practical part of it consists in doing, with regard to higher space, that which a plane-being would do with regard to our space in order to enable himself to realize what it was.

If we imagine one of these limited creatures whose life is cramped and confined studying the facts of space existence, we find that he can do it in two ways. He can assume another direction in addition to those which he knows; and he can, by means of abstract reasoning, say what would take place in an ampler kind of space than his own. All this would be formal work. The conclusions would be abstract possibilities.

The other mode of study is this. He can take some of these facts of his higher space and he can ponder over them in his mind, and can make up in his plane world those different appearances which one and the same solid body would present to him, and then he may try to realize inwardly what his higher existence is.

Now, it is evident that if the creature is absolutely confined to a two-dimensional existence, then anything more than such existence will always be a mere abstract and formal consideration to him.

But if this higher-space thought becomes real to him, if he finds in his mind a possibility of rising to it, then indeed he knows that somehow he is not limited to his apparent world. Everything he sees and comes into contact with may be two-dimensional; but essentially, somehow, himself he is not two-dimensional merely.

And a precisely similar piece of work is before us. Assuming as we must that our outer experience is limited to three-dimensional space, we shall make up the appearances which the very simplest higher bodies would present to us, and we shall gradually arrive at a more than merely formal and abstract appreciation of them. We shall discover in ourselves a faculty of apprehension of higher space similar to that which we have of space. And thus we shall discover, each for himself, that, limited as his senses are, he essentially somehow is not limited.

The mode and method in which this consciousness will be made general, is the same in which the spirit of an army is formed.

The individuals enter into the service from various motives, but each and all have to go through those movements and actions which correspond to the unity of a whole formed out of different members. The inner apprehension which lies in each man of a participation in a life wider than that of his individual body, is awakened and responds; and the active spirit of the army is formed. So with regard to higher space, this faculty of apprehending intuitively four-dimensional relationships will be taken up because of its practical use. Individuals will be practically employed to do it by society because of the larger faculty of thought which it gives. In fact, this higher-space thought means as an affair of mental training simply the power of apprehending the results arising from four independent causes. It means the power of dealing with a greater number of details.

And when this faculty of higher-space thought has been formed, then the faculty of apprehending that higher existence in which men have part, will come into being.

It is necessary to guard here against there being ascribed to this higher-space thought any other than an intellectual value. It has no moral value whatever. Its only connexion with moral or ethical considerations is the possibility it will afford of recognizing more of the facts of the universe than we do now. There is a gradual process going on which may be described as the getting rid of self elements. This process is one of knowledge and feeling, and either may be independent of the other. At present, in respect of feeling, we are much further on than in respect to understanding, and the reason is very much this: When a self element has been got rid of in respect of feeling, the new apprehension is put into practice, and we live it into our organization. But when a self element has been got rid of intellectually, it is allowed to remain a matter of theory, not vitally entering into the mental structure of individuals.

Thus up and down was discovered to be a self element more than a thousand years ago; but, except as a matter of theory, we are perfect barbarians in this respect up to the present day.

We have supposed a being living in a plane world, that is, a being of a very small thickness in a direction perpendicular to the surface on which he is.

Now, if we are situated analogously with regard to an ampler space, there must be some element in our experience corresponding to each element in the plane-being's experience.

And it is interesting to ask, in the case of the plane-being, what his opinion would be with respect to the surface on which he was.

He would not recognize it as a surface with which he was in contact; he would have no idea of a motion away from it or towards it.

But he would discover its existence by the fact that movements were transmitted along it. By its vibrating and quivering, it would impart movement to the particles of matter lying on it.

Hence, he would consider this surface to be a medium lying between bodies, and penetrating them. It would appear to him to have no weight, but to be a powerful means of transmitting vibrations. Moreover, it would be unlike any other substance with which he was acquainted, inasmuch as he could never get rid of it. However perfect a vacuum be made, there would be in this vacuum just as much of this unknown medium as there was before.

Moreover, this surface would not hinder the movement of the particles of matter over it. Being smooth, matter would slide freely over it. And this would seem to him as if matter went freely through the medium.

Then he would also notice the fact that vibrations of this medium would tear asunder portions of matter. The plane surface, being very compact, compared to the masses of matter on it, would, by its vibrations, shake them into their component parts.

Hence he would have a series of observations which tended to show that this medium was unlike any ordinary matter with which he was acquainted. Although matter passed freely through it, still by its shaking it could tear matter in pieces. These would be very difficult properties to reconcile in one and the same substance. Then it is weightless, and it is everywhere.

It might well be that he would regard the supposition of there being a plane surface, on which he was, as a preferable one to the hypothesis of this curious medium; and thus he might obtain a proof of his limitations from his observations.

Now, is there anything in our experience which corresponds to this medium which the plane-being gets to observe?

Do we suppose the existence of any medium through which matter freely moves, which yet by its vibrations destroys the combinations of matter—some medium which is present in every vacuum, however perfect, which penetrates all bodies, and yet can never be laid hold of?

These are precisely observations which have been made.

The substance which possesses all these qualities is called the æther. And the properties of the æther are a perpetual object of investigation in science.

Now, it is not the place here to go into details, as all we want is a basis for work; and however arbitrary it may be, it will serve if it enables us to investigate the properties of higher space.

We will suppose, then, that we are not in, but on the æther, only not on it in any known direction, but that the new direction is that which comes in. The æther is a smooth body, along which we slide, being distant from it at every point about the thickness of an atom; or, if we take our mean distance, being distant from it by half the thickness of an atom measured in this new direction.

Then, just as in space objects, a cube, for instance, can stand on the surface of a table, or on the surface over which the plane-being moves, so on the æther can stand a higher solid.

All that the plane-being sees or touches of a cube, is the square on which it rests.

So all that we could see or touch of a higher solid would be that part by which it stood on the æther; and this part would be to us exactly like any ordinary solid body. The base of a cube would be to the plane-being like a square which is to him an ordinary solid.

Now, the two ways, in which a plane-being would apprehend a solid body, would be by the successive appearances to him of it as it passed through his plane; and also by the different views of one and the same solid body which he got by turning the body over, so that different parts of its surface come into contact with his plane.

And the practical work of learning to think in four-dimensional space, is to go through the appearances which one and the same higher solid has.

Often, in the course of investigation in nature, we come across objects which have a certain similarity, and yet which are in parts entirely different. The work of the mind consists in forming an idea of that whole in which they cohere, and of which they are simple presentations.

The work of forming an idea of a higher solid is the most simple and most definite of all such mental operations.

If we imagine a plane world in which there are objects which correspond to our sun, to the planets, and, in fact, to all our visible universe, we must suppose a surface of enormous extent on which great disks slide, these disks being worlds of various orders of magnitude.

These disks would some of them be central, and hot, like our sun; round them would circulate other disks, like our planets.

And the systems of sun and planets must be conceived as moving with great velocity over the surface which bears them all.

And the movements of the atoms of these worlds will be the course of events in such worlds. As the atoms weave together, and form bodies altering, becoming, and ceasing, so will bodies be formed and disappear.

And the plane which bears them all on its smooth surface will simply be a support to all these movements, and influence them in no way.

Is to be conscious of being conscious of being hot, the same thing as to be conscious of being hot? It is not the same. There is a standing outside, and objectivation of a state of mind which every one would say in the first state was very different from the simple consciousness. But the consciousness must do as much in the first case as in the second. Hence the feeling hot is very different from the consciousness of feeling hot.

A feeling which we always have, we should not be conscious of—a sound always present ceases to be heard. Hence consciousness is a concomitant of change, that is, of the contact between one state and another.

If a being living on such a plane were to investigate the properties, he would have to suppose the solid to pass through his plane in order to see the whole of its surface. Thus we may imagine a cube resting on a table to begin to penetrate through the table. If the cube passes through the surface, making a clean cut all round it, so that the plane-being can come up to it and investigate it, then the different parts of the cube as it passes through the plane will be to him squares, which he apprehends by the boundary lines. The cut which there is in his plane must be supposed not to be noticed, he must be able to go right up to the cube without hindrance, and to touch and see that thin slice of it which is just above the plane.

And so, when we study a higher solid, we must suppose that it passes through the æther, and that we only see that thin three-dimensional section of it which is just about to pass from one side to the other of the æther.

When we look on a solid as a section of a higher solid, we have to suppose the æther broken through, only we must suppose that it runs up to the edge of the body which is penetrating it, so that we are aware of no breach of continuity.

The surface of the æther must then be supposed to have the properties of the surface of a fluid; only, of course, it is a solid three-dimensional surface, not a two-dimensional surface.

-

# CHAPTER IX.
# ANOTHER VIEW OF THE ÆTHER. MATERIAL AND ÆTHERIAL BODIES.

We have supposed in the case of a plane world that the surface on which the movements take place is inactive, except by its vibrations. It is simply a smooth support.

For the sake of simplicity let us call this smooth surface "the æther" in the case of a plane world.

The æther then we have imagined to be simply a smooth, thin sheet, not possessed of any definite structure, but excited by real disturbances of the matter on it into vibrations, which carry the effect of these disturbances as light and heat to other portions of matter. Now, it is possible to take an entirely different view of the æther in the case of a plane world.

Let us imagine that, instead of the æther being a smooth sheet serving simply as a support, it is definitely marked and grooved.

Let us imagine these grooves and channels to be very minute, but to be definite and permanent.

Then, let us suppose that, instead of the matter which slides in the æther having attractions and repulsions of its own, that it is quite inert, and has only the properties of inertia.

That is to say, taking a disk or a plane world as a specimen, the whole disk is sliding on the æther in virtue of a certain momentum which it has, and certain portions of its matter fit into the grooves in the æther, and move along those grooves.

The size of the portions is determined by the size of the grooves. And let us call those portions of matter which occupy the breadth of a groove, atoms. Then it is evident that the disk sliding along over the æther, its atoms will move according to the arrangement of the grooves over which the disk slides. If the grooves at any one particular place come close together, there will be a condensation of matter at that place when the disk passes over it; and if the grooves separate, there will be a rarefaction of matter.

If we imagine five particles, each slipping along in its own groove, if the particles are arranged in the form of a regular pentagon, and the grooves are parallel, then these five particles, moving evenly on, will maintain their positions with regard to one another, and a body would exist like a pentagon, lasting as long as the groves remained parallel.

But if, after some distance had been traversed by the disk, and these five particles were brought into a region where one of the grooves tended away from the others, the shape of the pentagon would be destroyed, it would become some irregular figure. And it is easy to see that if the grooves separated, and other grooves came in amongst them, along which other portions of matter were sliding, that the pentagon would disappear as an isolated body, that its constituent matter would be separated, and that its particles would enter into other shapes as constituents of them, and not of the original pentagon.

Thus, in cases of greater complication, an elaborate structure may be supposed to be formed, to alter, and to pass away; its origin, growth, and decay being due, not to any independent motion of the particles constituting it, but to the movement of the disk whereby its portions of matter were brought to regions where there was a particular disposition of the grooves.

Then the nature of the shape would really be determined by the grooves, not by the portions of matter which passed over them—they would become manifest as giving rise to a material form when a disk passed over them, but they would subsist independently of the disk; and if another disk were to pass over the same grooves, exactly the same material structures would spring up as came into being before.

If we make a similar supposition about our æther along which our earth slides, we may conceive the movements of the particles of matter to be determined, not by attractions or repulsions exerted on one another, but to be set in existence by the alterations in the directions of the grooves of the æther along which they are proceeding.

If the grooves were all parallel, the earth would proceed without any other motion than that of its path in the heavens.

But with an alteration in the direction of the grooves, the particles, instead of proceeding uniformly with the mass of the earth, would begin to move amongst each other. And by a sufficiently complicated arrangement of grooves it may be supposed that all the movements of the forms we see around us are due to interweaving and variously disposed grooves.

Thus the movements, which any body goes through, would depend on the arrangement of the æthereal grooves along which it was passing. As long as the grooves remain grouped together in approximately the same way, it would maintain its existence as the same body; but when the grooves separated, and became involved with the grooves of other objects, this body would cease to exist separately.

Thus the separate existences of the earth might conceivably be due to the disposition of those parts of the æther over which the earth passed. And thus any object would have to be separated into two parts, one the æthereal form, or modification which lasted, the other the material particles which, coming on with blind momentum, were directed into such movements as to produce the actual objects around us.

In this way there would be two parts in any organism, the material part and the æthereal part. There would be the material body, which soon passes and becomes indistinguishable from any other material body, and the æthereal body which remains.

Now, if we direct our attention to the material body, we see the phenomena of growth, decay, and death, the coming and the passing away of a living being, isolated during his existence, absolutely merged at his death into the common storehouse of matter.

But if we regard the æthereal body, we find something different. We find an organism which is not so absolutely separated from the surrounding organisms—an organism which is part of the æther, and which is linked to other æthereal organisms by its very substance—an organism between which and others there exists a unity incapable of being broken, and a common life which is rather marked than revealed by the matter which passes over it. The æthereal body moreover remains permanently when the material body has passed away.

The correspondences between the æthereal body and the life of an organism such as we know, is rather to be found in the emotional region than in the one of outward observation. To the æthereal form, all parts of it are equally one; but part of this form corresponds to the future of the material being, part of it to his past. Thus, care for the future and regard for the past would be the way in which the material being would exhibit the unity of the æthereal body, which is both his past, his present, and his future. That is to say, suppose the æthereal body capable of receiving an injury, an injury in one part of it would correspond to an injury in a man's past; an injury in another part,—that which the material body was traversing,—would correspond to an injury to the man at the present moment; injury to the æthereal body at another part, would correspond to injury coming to the man at some future time. And the self-preservation of the æthereal body, supposing it to have such a motive, would in the last case be the motive of regarding his own future to the man. And inasmuch as the man felt the real unity of his æthereal body, and did not confine his attention to his material body, which is absolutely disunited at every moment from its future and its past—inasmuch as he apprehended his æthereal unity, insomuch would he care for his future welfare, and consider it as equal in importance to his present comfort. The correspondence between emotion and physical fact would be, that the emotion of regard corresponded to an undiscerned æthereal unity. And then also, just as the two tips of two fingers put down on a plane, would seem to a plane-being to be two completely different bodies, not connected together, so one and the same æthereal body might appear as two distinct material bodies, and any regard between the two would correspond to an apprehension of their æthereal unity. In the supposition of an æthereal body, it is not necessary to keep to the idea of the rigidity and permanence of the grooves defining the motion of the matter which, passing along, exhibits the material body. The æthereal body may have a life of its own, relations with other æthereal bodies, and a life as full of vicissitudes as that of the material body, which in its total orbit expresses in the movements of matter one phase in the life of the æthereal body.

But there are certain obvious considerations which prevent any serious dwelling on these speculations—they are only introduced here in order to show how the conception of higher space lends itself to the representation of certain indefinite apprehensions,—such as that of the essential unity of the race,—and affords a possible clue to correspondences between the emotional and the physical life.

The whole question of our relation to the æther has to be settled. That which we call the æther is far more probably the surface of a liquid, and the phenomena we observe due to surface tensions. Indeed, the physical questions concern us here nothing at all. It is easy enough to make some supposition which gives us a standing ground to discipline our higher-space perception; and when that is trained, we shall turn round and look at the facts.

The conception which we shall form of the universe will undoubtedly be as different from our present one, as the Copernican view differs from the more pleasant view of a wide immovable earth beneath a vast vault. Indeed, any conception of our place in the universe will be more agreeable than the thought of being on a spinning ball, kicked into space without any means of communication with any other inhabitants of the universe.

# CHAPTER X.
# HIGHER SPACE AND HIGHER BEING. PERCEPTION AND INSPIRATION.

In the instinctive and sense perception of man and nature there is all hidden, which reflection afterwards brings into consciousness.

We are conscious of somewhat higher than each individual man when we look at men. In some, this consciousness reaches an extreme pitch, and becomes a religious apprehension. But in none is it otherwise than instinctive. The apprehension is sufficiently definite to be certain. But it is not expressible to us in terms of the reason.

Now, I have shown that by using the conception of higher space it is easy enough to make a supposition which shall show all mankind as physical parts of one whole. Our apparent isolation as bodies from each other is by no means so necessary to assume as it would appear. But, of course, a supposition of that kind is of no value, except as showing a possibility. If we came to examine into the matter closely, we should find a natural relationship which accounted for our consciousness being limited as at present it is.

The first thing to be done, is to organize our higher-space perception, and then look. We cannot tell what external objects will blend together into the unity of a higher being. But just as the riddle of the two hands becomes clear to us from our first inspection of higher space, so will there grow before our eyes greater unities and greater surprises.

We have been subject to a limitation of the most absurd character. Let us open our eyes and see the facts.

Now, it requires some training to open the eyes. For many years I worked at the subject without the slightest success. All was mere formalism. But by adopting the simplest means, and by a more thorough knowledge of space, the whole flashed clear.

Space shapes can only be symbolical of four-dimensional shapes; and if we do not deal with space shapes directly, but only treat them by symbols on the plane—as in analytical geometry—we are trying to get a perception of higher space through symbols of symbols, and the task is hopeless. But a direct study of space leads us to the knowledge of higher space. And with the knowledge of higher space there come into our ken boundless possibilities. All those things may be real, whereof saints and philosophers have dreamed.

Looking on the fact of life, it has become clear to the human mind, that justice, truth, purity, are to be sought—that they are principles which it is well to serve. And men have invented an abstract devotion to these, and all comes together in the grand but vague conception of Duty.

But all these thoughts are to those which spring up before us as the shadow on a bank of clouds of a great mountain is to the mountain itself. On the piled-up clouds falls the shadow—vast, imposing, but dark, colourless. If the beholder but turns, he beholds the mountain itself, towering grandly with verdant pines, the snowline, and the awful peaks.

So all these conceptions are the way in which now, with vision confined, we apprehend the great existences of the universe. Instead of an abstraction, what we have to serve is a reality, to which even our real things are but shadows. We are parts of a great being, in whose service, and with whose love, the utmost demands of duty are satisfied.

How can it not be a struggle, when the claims of righteousness mean diminished life,—even death,—to the individual who strives? And yet to a clear and more rational view it will be seen that in his extinction and loss, that which he loves,—that real being which is to him shadowed forth in the present existence of wife and child,—that being lives more truly, and in its life those he loves are his for ever.

But, of course, there are mistakes in what we consider to be our duty, as in everything else; and this is an additional reason for pursuing the quest of this reality. For by the rational observance of other material bodies than our own, we come to the conclusion that there are other beings around us like ourselves, whom we apprehend in virtue of two processes—the one simply a sense one of observation and reflection—the other a process of direct apprehension.

Now, if we did not go through the sense process of observation, we might, it is true, know that there were other human beings around us in some subtle way—in some mesmeric feeling; but we should not have that organized human life which, dealing with the things of the world, grows into such complicated forms. We should for ever be good-humoured babies—a sensuous, affectionate kind of jelly-fish.

And just so now with reference to the high intelligences by whom we are surrounded. We feel them, but we do not realize them.

To realize them, it will be necessary to develop our power of perception.

The power of seeing with our bodily eye is limited to the three-dimensional section.

But I have shown that the inner eye is not thus limited; that we can organize our power of seeing in higher space, and that we can form conceptions of realities in this higher space, just as we can in our ordinary space.

And this affords the groundwork for the perception and study of these other beings than man. Just as some mechanical means are necessary for the apprehension of our fellows in space, so a certain amount of mechanical education is necessary for the perception of higher beings in higher space.

Let us turn the current of our thought right round; instead of seeking after abstractions, and connecting our observations by ideas, let us train our sense of higher space and build up conceptions of greater realities, more absolute existences.

It is really a waste of time to write or read more generalities. Here is the grammar of the knowledge of higher being—let us learn it, not spend time in speculating as to whither it will lead us.

Yet one thing more. We are, with reference to the higher things of life, like blind and puzzled children. We know that we are members of one body, limbs of one vine; but we cannot discern, except by instinct and feeling, what that body is, what the vine is. If to know it would take away our feeling, then it were well never to know it. But fuller knowledge of other human beings does not take away our love for them; what reason is there then to suppose that a knowledge of the higher existences would deaden our feelings?

And then, again, we each of us have a feeling that we ourselves have a right to exist. We demand our own perpetuation. No man, I believe, is capable of sacrificing his life to any abstract idea; in all cases it is the consciousness of contact with some being that enables him to make the last human sacrifice. And what we can do by this study of higher space, is to make this consciousness, which has been reserved for a few, the property of all. Do we not all feel that there is a limit to our devotion to abstractions, none to beings whom we love. And to love them, we must know them.

Then, just as our own individual life is empty and meaningless without those we love, so the life of the human race is empty and meaningless without a knowledge of those that surround it. And although to some an inner knowledge of the oneness of all men is vouchsafed, it remains to be demonstrated to the many.

The perpetual struggle between individual interests and the common good can only be solved by merging both impulses in a love towards one being whose life lies in the fulfilment of each.

And this search, it seems to me, affords the needful supplement to the inquiries of one with whose thought I have been very familiar, and to which I return again, after having abandoned it for the purely materialistic views which seem forced upon us by the facts of science.

All that he said seemed to me unsupported by fact, unrelated to what we know.

But when I found that my knowledge was merely an empty pretence, that it was the vanity of being able to predict and foretell that stood to me in the place of an absolute apprehension of fact—when all my intellectual possessions turned to nothingness, then I was forced into that simple quest for fact, which, when persisted in and lived in, opens out to the thoughts like a flower to the life-giving sun.

It is indeed a far safer course, to believe that which appeals to us as noble, than simply to ask what is true; to take that which great minds have given, than to demand that our puny ones should be satisfied. But I suppose there is some good to some one in the scepticism and struggle of those who cannot follow in the safer course.

The thoughts of the inquirer to whom I allude may roughly be stated thus:—

He saw in human life the working out of a great process, in the toil and strain of our human history he saw the becoming of man. There is a defect whereby we fall short of the true measure of our being, and that defect is made good in the course of history.

It is owing to that defect that we perceive evil; and in the perception of evil and suffering lies our healing, for we shall be forced into that path at last, after trying every other, which is the true one.

And this, the history of the redemption of man, is what he saw in all the scenes of life; each most trivial occurrence was great and significant in relation to this.

And, further, he put forward a definite statement with regard to this defect, this lack of true being, for it lay, he said, in the self-centredness of our emotions, in the limitation of them to our bodily selves. He looked for a time when, driven from all thoughts of our own pain or pleasure, good or evil, we should say, in view of the miseries of our fellow-creatures, Let me be anyhow, use my body and my mind in any way, so that I serve.

And this, it seems to me, is the true aspiration; for, just as a note of music flings itself into the march of the melody, and, losing itself in it, is used for it and lost as a separate being, so we should throw these lives of ours as freely into the service of—whom?

Here comes the difficulty. Let it be granted that we should have no self-rights, limit our service in no way, still the question comes, What shall we serve?

It is far happier to have some concrete object to which we are devoted, or to be bound up in the ceaseless round of active life, wherein each day presents so many necessities that we have no room for choice.

But besides and apart from all these, there comes to some the question, "What does it all mean?" To others, an unlovable and gloomy aspect is presented, wherein their life seems to be but used as a material worthless in itself and ungifted with any dignity or honour; while to others again, with the love of those they love, comes a cessation of all personal interest in life, and a disappointment and feeling of valuelessness.

And in all these cases some answer is needed. And here human duty ceases. We cannot make objects to love. We can make machines and works of art, but nothing which directly excites our love. To give us that which rouses our love, is the duty of one higher than ourselves.

And yet in one respect we have a duty—we must look.

What good would it be, to surround us with objects of loving interest, if we bury our regards in ourselves and will not see?

And does it not seem as if with lowered eyelids, till only the thinnest slit was open, we gazed persistently, not on what is, but on the thinnest conceivable section of it?

Let it be granted that our right attitude is, so to devote ourselves that there is no question as to what we will do or what we will not do, but we are perfectly obedient servants. The question is, Whom are we to serve?

It cannot be each individual, for their claims are conflicting, and as often as not there is more need of a master than of a servant. Moreover, the aspect of our fellows does not always excite love, which is the only possible inducer of the right attitude of service. If we do not love, we can only serve for a self motive, because it is in some way good for ourselves.

Thus it seems to me that we are reduced to this: our only duty is to look for that which it is given us to love.

But this looking is not mere gazing. To know, we must act.

Let any one try it. He will find that unless he goes through a series of actions corresponding to his knowledge, he gets merely a theoretic and outside view of any facts. The way to know is this: Get somehow a means of telling what your perceptions would be if you knew, and act in accordance with those perceptions.

Thus, with regard to a fellow-creature, if we knew him we should feel what his feelings are. Let us then learn his feelings, and act as if we had them. It is by the practical work of satisfying his needs that we get to know him.

Then, may-be, we love him; or perchance it is said we may find that through him we have been brought into contact with one greater than him.

This is our duty—to know—to know, not merely theoretically, but practically; and then, when we know, we have done our part; if there is nothing, we cannot supply it. All we have to do is to look for realities.

We must not take this view of education—that we are horribly pressed for time, and must learn, somehow, a knack of saying how things must be, without looking at them.

But rather, we must say that we have a long time—all our lives, in which we will press facts closer and closer to our minds; and we begin by learning the simplest. There is an idea in that home of our inspiration—the fact that there are certain mechanical processes by which men can acquire merit. This is perfectly true. It is by mechanical processes that we become different; and the science of education consists largely in systematizing these processes.

Then, just as space perceptions are necessary for the knowledge of our fellow-men, and enable us to enter into human relationships with them in all the organized variety of civilized life, so it is necessary to develop our perceptions of higher space, so that we can apprehend with our minds the relationship which we have to beings higher than ourselves, and bring our instinctive knowledge into clearer consciousness.

It appears to me self-evident, that in the particular disposition of any portion of matter, that is, in any physical action, there can be neither right nor wrong; the thing done is perfectly indifferent.

At the same time, it is only in things done that we come into relationship with the beings about us and higher than us. Consequently, in the things we do lies the whole importance of our lives.

Now, many of our impulses are directly signs of a relationship in us to a being of which we are not immediately conscious. The feeling of love, for instance, is always directed towards a particular individual; but by love man tends towards the preservation and improvement of his race; thus in the commonest and most universal impulses lie his relations to higher beings than the individuals by whom he is surrounded. Now, along with these impulses are many instincts of a modifying tendency; and, being altogether in the dark as to the nature of the higher beings to whom we are related, it is difficult to say in what the service of the higher beings consists, in what it does not. The only way is, as in every other pre-rational department of life, to take the verdict of those with the most insight and inspiration.

And any striving against such verdicts, and discontent with them, should be turned into energy towards finding out exactly what relation we have towards these higher beings by the study of Space.

Human life at present is an art constructed in its regulations and rules on the inspirations of those who love the undiscerned higher beings, of which we are a part. They love these higher beings, and know their service.

But our perceptions are coarser; and it is only by labour and toil that we shall be brought also to see, and then lose the restraints that now are necessary to us in the fulness of love.

Exactly what relationship there is towards us on the part of these higher beings we cannot say in the least. We cannot even say whether there is more than humanity before the highest; and any conception which we form now must use the human drama as its only possible mode of presentation.

But that there is such a relation seems clear; and the ludicrous manner, in which our perceptions have been limited, is a sufficient explanation of why they have not been scientifically apprehended.

The mode, in which an apprehension of these higher beings or being is at present secured, is as follows; and it bears a striking analogy to the mode by which the self is cut out of a block of cubes.

When we study a block of cubes, we first of all learn it, by starting from a particular cube, and learning how all the others come with regard to that. All the others are right or left, up or down, near or far, with regard to that particular cube. And the line of cubes starting from this first one, which we take as the direction in which we look, is, as it were, an axis about which the rest of the cubes are grouped. We learn the block with regard to this axis, so that we can mentally conceive the disposition of every cube as it comes regarded from one point of view. Next we suppose ourselves to be in another cube at the extremity of another axis; and, looking from this axis, we learn the aspects of all the cubes, and so on.

Thus we impress on the feeling what the block of cubes is like from every axis. In this way we get a knowledge of the block of cubes.

Now, to get a knowledge of humanity, we must feel with many individuals. Each individual is an axis as it were, and we must regard human beings from many different axes. And as, in learning the block of cubes, muscular action, as used in putting up the block of cubes, is the means by which we impress on the feeling the different views of the block; so, with regard to humanity, it is by acting with regard to the view of each individual that a knowledge is obtained. That is to say, that, besides sympathizing with each individual, we must act with regard to his view; and acting so, we shall feel his view, and thus get to know humanity from more than one axis. Thus there springs up a feeling of humanity, and of more.

Those who feel superficially with a great many people, are like those learners who have a slight acquaintance with a block of cubes from many points of view. Those who have some deep attachments, are like those who know them well from one or two points of view.

Thus there are two definite paths—one by which the instinctive feeling is called out and developed, the other by which we gain the faculty of rationally apprehending and learning the higher beings.

In the one way it is by the exercise of a sympathetic and active life; in the other, by the study of higher space.

Both should be followed; but the latter way is more accessible to those who are not good. For we at any rate have the industry to go through mechanical operations, and know that we need something.

And after all, perhaps, the difference between the good and the rest of us, lies rather in the former being aware. There is something outside them which draws them to it, which they see while we do not.

There is no reason, however, why this knowledge should not become demonstrable fact. Surely, it is only by becoming demonstrable fact that the errors which have been necessarily introduced into it by human weakness will fall away from it.

The rational knowledge will not replace feeling, but will form the vehicle by which the facts will be presented to our consciousness. Just as we learn to know our fellows by watching their deeds,—but it is something beyond the mere power of observing them that makes us regard them,—so the higher existences need to be known; and, when known, then there is a chance that in the depths of our nature they will awaken feelings towards them like the natural response of one human being to another.

And when we reflect on what surrounds us, when we think that the beauty of fruit and flower, the blue depths of the sky, the majesty of rock and ocean,—all these are but the chance and arbitrary view which we have of true being,—then we can imagine somewhat of the glories that await our coming. How set out in exquisite loveliness are all the budding trees and hedgerows on a spring day—from here, where they almost sing to us in their nearness, to where, in the distance, they stand up delicately distant and distinct in the amethyst ocean of the air! And there, quiet and stately, revolve the slow moving sun and the stars of the night. All these are the fragmentary views which we have of great beings to whom we are related, to whom we are linked, did we but realize it, by a bond of love and service in close connexions of mutual helpfulness.

Just as here and there on the face of a woman sits the divine spirit of beauty, so that all cannot but love who look—so, presenting itself to us in all this mingled scene of air and ocean, plain and mountain, is a being of such loveliness that, did we but know with one accord in one stream, all our hearts would be carried in a perfect and willing service. It is not that we need to be made different; we have but to look and gaze, and see that centre whereunto with joyful love all created beings move.

But not with effortless wonder will our days be filled, but in toil and strong exertion; for, just as now we all labour and strive for an object, our service is bound up with things which we do—so then we find no rest from labour, but the sense of solitude and isolation is gone. The bonds of brotherhood with our fellow-men grow strong, for we know one common purpose. And through the exquisite face of nature shines the spiritual light that gives us a great and never-failing comrade.

Our task is a simple one—to lift from our mind that veil which somehow has fallen on us, to take that curious limitation from our perception, which at present is only transcended by inspiration.

And the means to do it is by throwing aside our reason—by giving up the idea that what we think or are has any value. We too often sit as judges of nature, when all we can be are her humble learners. We have but to drink in of the inexhaustible fulness of being, pressing it close into our minds, and letting our pride of being able to foretell vanish into dust.

There is a curious passage in the works of Immanuel Kant,[1] in which he shows that space must be in the mind before we can observe things in space. "For," he says, "since everything we conceive is conceived as being in space, there is nothing which comes before our minds from which the idea of space can be derived; it is equally present in the most rudimentary perception and the most complete." Hence he says that space belongs to the perceiving soul itself. Without going into this argument to abstract regions, it has a great amount of practical truth. All our perceptions are of things in space; we cannot think of any detail, however limited or isolated, which is not in space.

Hence, in order to exercise our perceptive powers, it is well to have prepared beforehand a strong apprehension of space and space relations.

And so, as we pass on, is it not easily conceivable that, with our power of higher space perception so rudimentary and so unorganized, we should find it impossible to perceive higher existences? That mode of perception which it belongs to us to exercise is wanting. What wonder, then, that we cannot see the objects which are ready, were but our own part done?

Think how much has come into human life through exercising the power of the three-dimensional space perception, and we can form some measure, in a faint way, of what is in store for us.

There is a certain reluctance in us in bringing anything, which before has been a matter of feeling, within the domain of conscious reason. We do not like to explain why the grass is green, flowers bright, and, above all, why we have the feelings which we pass through.

But this objection and instinctive reluctance is chiefly derived from the fact that explaining has got to mean explaining away. We so often think that a thing is explained, when it can be shown simply to be another form of something which we know already. And, in fact, the wearied mind often does long to have a phenomenon shown to be merely a deduction from certain known laws.

But explanation proper is not of this kind; it is introducing into the mind the new conception which is indicated by the phenomenon already present. Nature consists of many entities towards the apprehension of which we strive. If for a time we break down the bounds which we have set up, and unify vast fields of observation under one common law, it is that the conceptions we formed at first are inadequate, and must be replaced by greater ones. But it is always the case, that, to understand nature, a conception must be formed in the mind. This process of growth in the mental history is hidden; but it is the really important one. The new conception satisfies more facts than the old ones, is truer phenomenally; and the arguments for it are its simplicity, its power of accounting for many facts. But the conception has to be formed first. And the real history of advance lies in the growth of the new conceptions which every now and then come to light.

When the weather-wise savage looked at the sky at night, he saw many specks of yellow light, like fire-flies, sprinkled amidst whitish fleece; and sometimes the fleece remained, the fire-spots went, and rain came; sometimes the fire-spots remained, and the night was fine. He did not see that the fire-points were ever the same, the clouds different; but by feeling dimly, he knew enough for his purpose.

But when the thinking mind turned itself on these appearances, there sprang up,—not all at once, but gradually,—the knowledge of the sublime existences of the distant heavens, and all the lore of the marvellous forms of water, of air, and the movements of the earth. Surely these realities, in which lies a wealth of embodied poetry, are well worth the delighted sensuous apprehension of the savage as he gazed.

Perhaps something is lost, but in the realities, of which we know, there is compensation. And so, when we learn to understand the meaning of these mysterious changes, this course of natural events, we shall find in the greater realities amongst which we move a fair exchange for the instinctive reverence, which they now awaken in us.

In this book the task is taken up of forming the most simple and elementary of the great conceptions that are about us. In the works of the poets, and still more in the pages of religious thinkers, lies an untold wealth of conception, the organization of which in our every-day intellectual life is the work of the practical educator.

But none is capable of such simple demonstration and absolute presentation as this of higher space, and none so immediately opens our eyes to see the world as a different place. And, indeed, it is very instructive; for when the new conception is formed, it is found to be quite simple and natural. We ask ourselves what we have gained; and we answer: Nothing; we have simply removed an obvious limitation.

And this is universally true; it is not that we must rise to the higher by a long and laborious process. We may have a long and laborious process to go through, but, when we find the higher, it is this: we discover our true selves, our essential being, the fact of our lives. In this case, we pass from the ridiculous limitation, to which our eyes and hands seem to be subject, of acting in a mere section of space, to the fuller knowledge and feeling of space as it is. How do we pass to this truer intellectual life? Simply by observing, by laying aside our intellectual powers, and by looking at what is.

We take that which is easiest to observe, not that which is easiest to define; we take that which is the most definitely limited real thing, and use it as our touchstone whereby to explore nature.

As it seems to me, Kant made the great and fundamental statement in philosophy when he exploded all previous systems, and all physics were reft from off the perceiving soul. But what he did once and for all, was too great to be a practical means of intellectual work. The dynamic form of his absolute insight had to be found; and it is in other works that the practical instances of the Kantian method are to be found. For, instead of looking at the large foundations of knowledge, the ultimate principles of experience, late writers turned to the details of experience, and tested every phenomenon, not with the question, What is this? but with the question, "What makes me perceive thus?"

And surely the question, as so put, is more capable of an answer; for it is only the percipient, as a subject of thought, about which we can speak. The absolute soul, since it is the thinker, can never be the subject of thought; but, as physically conditioned, it can be thought about. Thus we can never, without committing a ludicrous error, think of the mind of man except as a material organ of some kind; and the path of discovery lies in investigating what the devious line of his thought history is due to, which winds between two domains of physics—the unknown conditions which affect the perceiver, the partially known physics which constitute what we call the external world.

It is a pity to spend time over these reflections; if they do not seem tame and poor compared to the practical apprehension which comes of working with the models, then there is nothing in the whole subject. If in the little real objects which the reader has to handle and observe does not lie to him a poetry of a higher kind than any expressed thought, then all these words are not only useless, but false. If, on the other hand, there is true work to be done with them, then these suggestions will be felt to be but mean and insufficient apprehensions.

For, in the simplest apprehension of a higher space lies a knowledge of a reality which is, to the realities we know, as spirit is to matter; and yet to this new vision all our solid facts and material conditions are but as a shadow is to that which casts it. In the awakening light of this new apprehension, the flimsy world quivers and shakes, rigid solids flow and mingle, all our material limitations turn into graciousness, and the new field of possibility waits for us to look and behold.

# CHAPTER XI.
# SPACE THE SCIENTIFIC BASIS OF ALTRUISM AND RELIGION.

The reader will doubtless ask for some definite result corresponding to these words—something not of the nature of an hypothesis or a might-be. And in that I can only satisfy him after my own powers. My only strength is in detail and patience; and if he will go through the practical part of the book, it will assuredly dawn upon him that here is the beginning of an answer to his request. I only study the blocks and stones of the higher life. But here they are definite enough. And the more eager he is for personal and spiritual truth, the more eagerly do I urge him to take up the practical work, for the true good comes to us through those who, aspiring greatly, still submit their aspirations to fact, and who, desiring to apprehend spirit, still are willing to manipulate matter.

The particular problem at which I have worked for more than ten years, has been completely solved. It is possible for the mind to acquire a conception of higher space as adequate as that of our three-dimensional space, and to use it in the same manner.

There are two distinct ways of studying space—our familiar space at present in use. One is that of the analyst, who treats space relations by his algebra, and discovers marvellous relations. The other is that of the observer or mechanician, who studies the shapes of things in space directly.

A practical designer of machines would not find the knowledge of geometrical analysis of immediate help to him; and an artist or draughtsman still less so.

Now, my inquiry was, whether it was possible to get the same power of conception of four-dimensional space, as the designer and draughtsman have of three-dimensional space. It is possible.

And with this power it is possible for us to design machines in higher space, and to conceive objects in this space, just as a draughtsman or artist does.

Analytical skill is not of much use in designing a statue or inventing a machine, or in appreciating the detail of either a work of art or a mechanical contrivance.

And hitherto the study of four-dimensional space has been conducted by analysis. Here, for the first time, the fact of the power of conception of four-dimensional space is demonstrated, and the means of educating it are given.

And I propose a complete system of work, of which the volume on four space is the first instalment.

[2] "Science Romance," No. I., by C. H. Hinton. Published by Swan Sonnenschein & Co.

I shall bring forward a complete system of four-dimensional thought—mechanics, science, and art. The necessary condition is, that the mind acquire the power of using four-dimensional space as it now does three-dimensional.

And there is another condition which is no less important. We can never see, for instance, four-dimensional pictures with our bodily eyes, but we can with our mental and inner eye. The condition is, that we should acquire the power of mentally carrying a great number of details.

If, for instance, we could think of the human body right down to every minute part in its right position, and conceive its aspect, we should have a four-dimensional picture which is a solid structure. Now, to do this, we must form the habit of mental painting, that is, of putting definite colours in definite positions, not with our hands on paper, but with our minds in thought, so that we can recall, alter, and view complicated arrangements of colour existing in thought with the same ease with which we can paint on canvas. This is simply an affair of industry; and the mental power latent in us in this direction is simply marvellous.

In any picture, a stroke of the brush put on without thought is valueless. The artist is not conscious of the thought process he goes through. For our purpose it is necessary that the manipulation of colour and form which the artist goes through unconsciously, should become a conscious power, and that, at whatever sacrifice of immediate beauty, the art of mental painting should exist beside our more unconscious art. All that I mean is this—that in the course of our campaign it is necessary to take up the task of learning pictures by heart, so that, just as an artist thinks over the outlines of a figure he wants to draw, so we think over each stroke in our pictures. The means by which this can be done will be given in a future volume.

We throw ourselves on an enterprise in which we have to leave altogether the direct presentation to the senses. We must acquire a sense-perception and memory of so keen and accurate a kind that we can build up mental pictures of greater complexity than any which we can see. We have a vast work of organization, but it is merely organization. The power really exists and shows itself when it is looked for.

Much fault may be found with the system of organization which I have adopted, but it is the survivor of many attempts; and although I could better it in parts, still I think it is best to use it until, the full importance of the subject being realized, it will be the lifework of men of science to reorganize the methods.

The one thing on which I must insist is this—that knowledge is of no value, it does not exist unless it comes into the mind. To know that a thing must be is no use at all. It must be clearly realized, and in detail as it *is*, before it can be used.

A whole world swims before us, the apprehension of which simply demands a patient cultivation of our powers; and then, when the faculty is formed, we shall recognize what the universe in which we are is like. We shall learn about ourselves and pass into a new domain.

And I would speak to some minds who, like myself, share to a large extent the feeling of unsettledness and unfixedness of our present knowledge.

Religion has suffered in some respects from the inaccuracy of its statements; and it is not always seen that it consists of two parts—one a set of rules as to the management of our relations to the physical world about us, and to our own bodies; another, a set of rules as to our relationship to beings higher than ourselves.

Now, on the former of these subjects, on physical facts, on the laws of health, science has a fair standing ground of criticism, and can correct the religious doctrines in many important respects.

But on the other part of the subject matter, as to our relationship to beings higher than ourselves, science has not yet the materials for judging. The proposition which underlies this book is, that we should begin to acquire the faculties for judging.

To judge, we must first appreciate; and how far we are from appreciating with science the fundamental religious doctrines I leave to any one to judge.

There is absolutely no scientific basis for morality, using morality in the higher sense of other than a code of rules to promote the greatest physical and mental health and growth of a human being. Science does not give us any information which is not equally acceptable to the most selfish and most generous man; it simply tells him of means by which he may attain his own ends, it does not show him ends.

The prosecution of science is an ennobling pursuit; but it is of scientific knowledge that I am now speaking in itself. We have no scientific knowledge of any existences higher than ourselves—at least, not recognized as higher. But we have abundant knowledge of the actions of beings less developed than ourselves, from the striking unanimity with which all inorganic beings tend to move towards the earth's centre, to the almost equally uniform modes of response in elementary organized matter to different stimuli.

The question may be put: In what way do we come into contact with these higher beings at present? And evidently the answer is, In those ways in which we tend to form organic unions—unions in which the activities of individuals coalesce in a living way.

The coherence of a military empire or of a subjugated population, presenting no natural nucleus of growth, is not one through which we should hope to grow into direct contact with our higher destinies. But in friendship, in voluntary associations, and above all, in the family, we tend towards our greater life.

And it seems that the instincts of women are much more relative to this, the most fundamental and important side of life, than are those of men. In fact, until we know, the line of advance had better be left to the feeling of women, as they organize the home and the social life spreading out therefrom. It is difficult, perhaps, for a man to be still and perceive; but if he is so, he finds that what, when thwarted, are meaningless caprices and empty emotionalities, are, on the part of woman, when allowed to grow freely and unchecked, the first beginnings of a new life—the shadowy filaments, as it were, by which an organism begins to coagulate together from the medium in which it makes its appearance.

In very many respects men have to make the conditions, and then learn to recognize. How can we see the higher beings about us, when we cannot even conceive the simplest higher shapes? We may talk about space, and use big words, but, after all, the preferable way of putting our efforts is this: let us look first at the simplest facts of higher existence, and then, when we have learnt to realize these, We shall be able to see what the world presents. And then, also, light will be thrown on the constituent organisms of our own bodies, when we see in the thorough development of our social life a relation between ourselves and a larger organism similar to that which exists between us and the minute constituents of our frame.

The problem, as it comes to me, is this: it is clearly demonstrated that self-regard is to be put on one side—and self-regard in every respect—not only should things painful and arduous be done, but things degrading and vile, so that they serve.

I am to sign any list of any number of deeds which the most foul imagination can suggest, as things which I would do did the occasion come when I could benefit another by doing them; and, in fact, there is to be no characteristic in any action which I would shrink from did the occasion come when it presented itself to be done for another's sake. And I believe that the soul is absolutely unstained by the action, provided the regard is for another.

But this is, in truth, a dangerous doctrine; at one Sweep it puts away all absolute commandments, all absolute verdicts of right about things, and leaves the agent to his own judgment.

It is a kind of rule of life which requires most absolute openness, and demands that society should frame severe and insuperable regulations; for otherwise, with the motives of the individual thus liberated from absolute law, endless varieties of conduct would spring forth, and the wisdom of individual men is hardly enough to justify their irresponsible action.

Still, it does seem that, as an ideal, the absolute absence of self-regard is to be aimed at.

With a strong religious basis, this would work no harm, for the rules of life, as laid down by religions, would suffice. But there are many who do not accept these rules as any absolute indication of the will of God, but only as the regulations of good men, which have a claim to respect and nothing more.

And thus it seems to me that altruism—thoroughgoing altruism—hands over those who regard it as an ideal, and who are also of a sceptical turn of mind, to the most absolute unfixedness of theory, and, very possibly, to the greatest errors in life.

And here we come to the point where the study of space becomes so important.

For if this rule of altruism is the right one, if it appeals with a great invitation to us, we need not therefore try it with less precaution than we should use in other affairs of infinitely less importance. When we want to know if a plank will bear, we entrust it with a different load from that of a human body.

And if this law of altruism is the true one, let us try it where failure will not mean the ruin of human beings.

Now, in knowledge, pure altruism means so to bury the mind in the thing known that all particular relations of one's self pass away. The altruistic knowledge of the heavens would be, to feel that the stars were vast bodies, and that I am moving rapidly. It would be, to know this, not as a matter of theory, but as a matter of habitual feeling.

Whether this is possible, I do not know; but a somewhat similar attempt can be made with much simpler means.

In a different place I have described the process of acquiring an altruistic knowledge of a block of cubes; and the results of the laborious processes involved are well worth the trouble. For as a clearly demonstrable fact this comes before one. To acquire an absolute knowledge of a block of cubes, so that all self relations are cast out, means that one has to take the view of a higher being.

It suddenly comes before one, that the particular relations which are so fixed and important, and seem so absolutely sure when one begins the process of learning, are by no means absolute facts, but marks of a singular limitation, almost a degradation, on one's own part. In the determined attempt to know the most insignificant object perfectly and thoroughly, there flashes before one's eyes an existence infinitely higher than one's own. And with that vision there comes,—I do not speak from my own experience only,—a conviction that our existence also is not what we suppose—that this bodily self of ours is but a limit too. And the question of altruism, as against self-regard, seems almost to vanish, for by altruism we come to know what we truly are.

"What we truly are," I do not mean apart from space and matter, but what we really are as beings having a space existence; for our way of thinking about existence is to conceive it as the relations of bodies in space. To think is to conceive realities in space.

Just as, to explore the distant stars of the heavens, a particular material arrangement is necessary which we call a telescope, so to explore the nature of the beings who are higher than us, a mental arrangement is necessary. We must prepare our power of thinking as we prepare a more extended power of looking. We want a structure developed inside the skull for the one purpose, while an exterior telescope will do for the other.

And thus it seems that the difficulties which we first apprehended fall away.

To us, looking with half-blinded eyes at merely our own little slice of existence, our filmy all, it seemed that altruism meant disorder, vagary, danger.

But when we put it into practice in knowledge, we find that it means the direct revelation of a higher being and a call to us to participate ourselves too in a higher life—nay, a consciousness comes that we are higher than we know.

And so with our moral life as with our intellectual life. Is it not the case that those, who truly accept the rule of altruism, learn life in new dangerous ways?

It is true that we must give up the precepts of religion as being the will of God; but then we shall learn that the will of God shows itself partly in the religious precepts, and comes to be more fully and more plainly known as an inward spirit.

And that difficulty, too, about what we may do and what we may not, vanishes also. For, if it is the same about our fellow-creatures as it is about the block of cubes, when we have thrown out the self-regard from our relationship to them, we shall feel towards them as a higher being than man feels towards them, we shall feel towards them as they are in their true selves, not in their outward forms, but as eternal loving spirits.

And then those instincts which humanity feels with a secret impulse to be sacred and higher than any temporary good will be justified—or fulfilled.

There are two tendencies—one towards the direct cultivation of the religious perceptions, the other to reducing everything to reason. It will be but just for the exponents of the latter tendency to look at the whole universe, not the mere section of it which we know, before they deal authoritatively with the higher parts of religion.

And those who feel the immanence of a higher life in us will be needed in this outlook on the wider field of reality, so that they, being fitted to recognize, may tell us what lies ready for us to know.

The true path of wisdom consists in seeing that our intellect is foolishness—that our conclusions are absurd and mistaken, not in speculating on the world as a form of thought projected from the thinking principle within us—rather to be amazed that our thought has so limited the world and hidden from us its real existences. To think of ourselves as any other than things in space and subject to material conditions, is absurd, it is absurd on either of two hypotheses. If we are really things in space, then of course it is absurd to think of ourselves as if we were not so. On the other hand, if we are not things in space, then conceiving in space is the mode in which that unknown which we are exists as a mind. Its mental action is space-conception, and then to give up the idea of ourselves as in space, is not to get a truer idea, but to lose the only power of apprehension of ourselves which we possess.

And yet there is, it must be confessed, one way in which it may be possible for us to think without thinking of things in space.

That way is, not to abandon the use of space-thought, but to pass through it.

When we think of space, we have to think of it as infinity extended, and we have to think of it as of infinite dimensions. Now, as I have shown in "The Law of the Valley,"[3] when we come upon infinity in any mode of our thought, it is a sign that that mode of thought is dealing with a higher reality than it is adapted for, and in struggling to represent it, can only do so by an infinite number of terms. Now, space has an infinite number of positions and turns, and this may be due to the attempt forced upon us to think of things higher than space as in space. If so, then the way to get rid of space from our thoughts, is, not to go away from it, but to pass through it—to think about larger and larger systems of space, and space of more and more dimensions, till at last we get to such a representation in space of what is higher than space, that we can pass from the space-thought to the more absolute thought without that leap which would be necessary if we were to try to pass beyond space with our present very inadequate representation in it of what really is.

[3] "Science Romances," No. II.

Again and again has human nature aspired and fallen. The vision has presented itself of a law which was love, a duty which carried away the enthusiasm, and in which the conflict of the higher and lower natures ceased because all was enlisted in one loving service. But again and again have such attempts failed. The common-sense view, that man is subject to law, external law, remains—that there are fates whom he must propitiate and obey. And there is a strong sharp curb, which, if it be not brought to bear by the will, is soon pulled tight by the world, and one more tragedy is enacted, and the over-confident soul is brought low.

And the rock on which such attempts always split, is in the indulgence of some limited passion. Some one object fills the soul with its image, and in devotion to that, other things are sacrificed, until at last all comes to ruin.

But what does this mean? Surely it is simply this, that where there should be knowledge there is ignorance. It is not that there is too much devotion, too much passion, but that we are ignorant and blind, and wander in error. We do not know what it is we care for, and waste our effort on the appearance. There is no such thing as wrong love; there is good love and bad knowledge, and men who err, clasp phantoms to themselves. Religion is but the search for realities; and thought, conscious of its own limitations, is its best aid.

Let a man care for any one object—let his regard for it be as concentrated and exclusive as you will, there will be no danger if he truly apprehends that which he cares for. Its true being is bound up with all the rest of existence, and, if his regard is true to one, then, if that one is really known, his regard is true to all.

There is a question sometimes asked, which shows the mere formalism into which we have fallen.

We ask: What is the end of existence? A mere play on words! For to conceive existence is to feel ends. The knowledge of existence is the caring for objects, the fear of dangers, the anxieties of love. Immersed in these, the triviality of the question, what is the end of existence? becomes obvious. If, however, letting reality fade away, we play with words, some questions of this kind are possible; but they are mere questions of words, and all content and meaning has passed out of them.

The task before us is this: we strive to find out that physical unity, that body which men are parts of, and in the life of which their true unity lies. The existence of this one body we know from the utterances of those whom we cannot but feel to be inspired; we feel certain tendencies in ourselves which cannot be explained except by a supposition of this kind.

And, now, we set to work deliberately to form in our minds the means of investigation, the faculty of higher-space conception. To our ordinary space-thought, men are isolated, distinct, in great measure antagonistic. But with the first use of the weapon of higher thought, it is easily seen that all men may really be members of one body, their isolation may be but an affair of limited consciousness. There is, of course, no value as science in such a supposition. But it suggests to us many possibilities; it reveals to us the confined nature of our present physical views, and stimulates us to undertake the work necessary to enable us to deal adequately with the subject.

The work is entirely practical and detailed; it is the elaboration, beginning from the simplest objects of an experience in thought, of a higher-space world.

To begin it, we take up those details of position and relation which are generally relegated to symbolism or unconscious apprehension, and bring these waste products of thought into the central position of the laboratory of the mind. We turn all our attention on the most simple and obvious details of our every-day experience, and thence we build up a conception of the fundamental facts of position and arrangement in a higher world. We next study more complicated higher shapes, and get our space perception drilled and disciplined. Then we proceed to put a content into our framework.

The means of doing this are twofold—observation and inspiration.

As to observation, it is hardly possible to describe the feelings of that investigator who shall distinctly trace in the physical world, and experimentally demonstrate the existence of the higher-space facts which are so curiously hidden from us. He will lay the first stone for the observation and knowledge of the higher beings to whom we are related.

As to the other means, it is obvious, surely, that if there has ever been inspiration, there is inspiration now. Inspiration is not a unique phenomenon. It has existed in absolutely marvellous degree in some of the teachers of the ancient world; but that, whatever it was, which they possessed, must be present now, and, if we could isolate it, be a demonstrable fact.

And I would propose to define inspiration as the faculty, which, to take a particular instance, does the following:—

If a square penetrates a line cornerwise, it marks out on the line a segment bounded by two points—that is, we suppose a line drawn on a piece of paper, and a square lying on the paper to be pushed so that its corner passes over the line. Then, supposing the paper and the line to be in the same plane, the line is interrupted by the square; and, of the square, all that is observable in the line, is a segment bounded by two points.

Next, suppose a cube to be pushed cornerwise through a plane, and let the plane make a section of the cube. The section will be a plane figure, and it will be a triangle.

Now, first, the section of a square by a line is a segment bounded by two points; second, the section of a cube by a plane is a triangle bounded by three lines.

Hence, we infer that the section of a figure in four dimensions analogous to a cube, by three-dimensional space, will be a tetrahedron—a figure bounded by four planes.

This is found to be true; with a little familiarity with four-dimensional movements this is seen to be obvious. But I would define inspiration as the faculty by which without actual experience this conclusion is formed.

How it is we come to this conclusion I am perfectly unable to say. Somehow, looking at mere formal considerations, there comes into the mind a conclusion about something beyond the range of actual experience.

We may call this reasoning from analogy; but using this phrase does not explain the process. It seems to me just as rational to say that the facts of the line and plane remind us of facts which we know already about four-dimensional figures—that they tend to bring these facts out into consciousness, as Plato shows with the boy's knowledge of the cube. We must be really four-dimensional creatures, or we could not think about four dimensions.

But whatever name we give to this peculiar and inexplicable faculty, that we do possess it is certain; and in our investigations it will be of service to us. We must carefully investigate existence in a plane world, and then, making sure, and impressing on our inward sense, as we go, every step we take with regard to a higher world, we shall be reminded continually of fresh possibilities of our higher existence.

# PART II.

## CHAPTER I.
## THREE-SPACE. GENESIS OF A CUBE. APPEARANCES OF A CUBE TO A PLANE-BEING.

The models consist of a set of eight and a set of four cubes. They are marked with different colours, so as to show the properties of the figure in Higher Space, to which they belong.

The simplest figure in one-dimensional space, that is, in a straight line, is a straight line bounded at the two extremities. The figure in this case consists of a length bounded by two points.

Looking at Cube 1, and placing it so that the figure 1 is uppermost, we notice a straight line in contact with the table, which is coloured Orange. It begins in a Gold point and ends in a Fawn point. The Orange extends to some distance on two faces of the Cube; but for our present purpose we suppose it to be simply a thin line.

This line we conceive to be generated in the following way. Let a point move and trace out a line. Let the point be the Gold point, and let it, moving, trace out the Orange line and terminate in the Fawn point. Thus the figure consists of the point at which it begins, the point at which it ends, and the portion between. We may suppose the point to start as a Gold point, to change its colour to Orange during the motion, and when it stops to become Fawn. The motion we suppose from left to right, and its direction we call X.

If, now, this Orange line move away from us at right angles, it will trace out a square. Let this be the Black square, which is seen underneath Model 1. The points, which bound the line, will during this motion trace out lines, and to these lines there will be terminal points. Also, the Square will be terminated by a line on the opposite side. Let the Gold point in moving away trace out a Blue line and end in a Buff point; the Fawn point a Crimson line ending in a Terracotta point. The Orange line, having traced a Black square, ends in a Green-grey line. This direction, away from the observer, we call Y.

Now, let the whole Black square traced out by the Orange line move upwards at right angles. It will trace out a new figure, a Cube. And the edges of the square, while moving upwards, will trace out squares. Bounding the cube, and opposite to the Black square, will be another square. Let the Orange line moving upwards trace a Dark Blue square and end in a Reddish line. The Gold point traces a Brown line; the Fawn point traces a French-grey line, and these lines end in a Light-blue and a Dull-purple point. Let the Blue line trace a Vermilion square and end in a Deep-yellow line. Let the Buff point trace a Green line, and end in a Red point. The Green-grey line traces a Light-yellow square and ends in a Leaden line; the Terracotta point traces a Dark-slate line and ends in a Deep-blue point. The Crimson line traces a Blue-green square and ends in a Bright-blue line.

Finally, the Black square traces a Cube, the colour of which is invisible, and ends in a white square. We suppose the colour of the cube to be a Light-buff. The upward direction we call Z. Thus we say: The Gold point moved Z, traces a Brown line, and ends in a Light-blue point.

We can now clearly realize and refer to each region of the cube by a colour.

At the Gold point, lines from three directions meet, the X line Orange, the Y line Blue, the Z line Brown.

Thus we began with a figure of one dimension, a line, we passed on to a figure of two dimensions, a square, and ended with a figure of three dimensions, a cube.

-

The square represents a figure in two dimensions; but if we want to realize what it is to a being in two dimensions, we must not look down on it. Such a view could not be taken by a plane-being.

Let us suppose a being moving on the surface of the table and unable to rise from it. Let it not know that it is upon anything, but let it believe that the two directions and compounds of those two directions are all possible directions. Moreover, let it not ask the question: "On what am I supported?" Let it see no reason for any such question, but simply call the smooth surface, along which it moves, Space.

Such a being could not tell the colour of the square traced by the Orange line. The square would be bounded by the lines which surround it, and only by breaking through one of those lines could the plane-being discover the colour of the square.

In trying to realize the experience of a plane-being it is best to suppose that its two dimensions are upwards and sideways, *i.e.*, Z and X, because, if there be any matter in the plane-world, it will, like matter in the solid world, exert attractions and repulsions. The matter, like the beings, must be supposed very thin, that is, of so slight thickness that it is quite unnoticed by the being. Now, if there be a very large mass of such matter lying on the table, and a plane-being be free to move about it, he will be attracted to it in every direction. "Towards this huge mass" would be "Down," and "Away from it" would be "Up," just as "Towards the earth" is to solid beings

"Down," and "Away from it" is "Up," at whatever part of the globe they may be. Hence, if we want to realize a plane-being's feelings, we must keep the sense of up and down. Therefore we must use the Z direction, and it is more convenient to take Z and X than Z and Y.

Any direction lying between these is said to be compounded of the two; for, if we move slantwise for some distance, the point reached might have been also reached by going a certain distance X, and then a certain distance Z, or *vice versâ*.

Let us suppose the Orange line has moved Z, and traced the Dark-blue square ending in the Reddish line. If we now place a piece of stiff paper against the Dark-blue square, and suppose the plane-beings to move to and fro on that surface of the paper, which touches the square, we shall have means of representing their experience.

To obtain a more consistent view of their existence, let us suppose the piece of paper extended, so that it cuts through our earth and comes out at the antipodes, thus cutting the earth in two. Then suppose all the earth removed away, both hemispheres vanishing, and only a very thin layer of matter left upon the paper on that side which touches the Dark-blue square. This represents what the world would be to a plane-being.

It is of some importance to get the notion of the directions in a plane-world, as great difficulty arises from our notions of up and down. We miss the right analogy if we conceive of a plane-world without the conception of up and down.

A good plan is, to use a slanting surface, a stiff card or book cover, so placed that it slopes upwards to the eye. Then gravity acts as two forces. It acts (1) as a force pressing all particles upon the slanting surface into it, and (2) as a force of gravity along the plane, making particles tend to slip down its incline. We may suppose that in a plane-world there are two such forces, one keeping the beings thereon to the plane, the other acting between bodies in it, and of such a nature that by virtue of it any large mass of plane-matter produces on small particles around it the same effects as the large mass of solid matter called our earth produces on small objects like our bodies situated around it. In both cases the larger draws the smaller to itself, and creates the sensations of up and down.

If we hold the cube so that its Dark-blue side touches a sheet of paper held upwards to the eye, and if we then look straight down along the paper, confining our view to that which is in actual contact with the paper, we see the same view of the cube as a plane-being would get. We see a Light-blue point, a Reddish line, and a Dull-purple point. The plane-being only sees a line, just as we only see a square of the cube.

The line where the paper rests on the table may be taken as representative of the surface of the plane-being's earth. It would be merely a line to him, but it would have the same property in relation to the plane-world, as a square has in relation to a solid world; in neither case can the notion of what in the latter is termed solidity be quite excluded. If the plane-being broke through the line bounding his earth, he would find more matter beyond it.

Let us now leave out of consideration the question of "up and down" in a plane-world. Let us no longer consider it in the vertical, or ZX, position, but simply take the surface (XY) of the table as that which supports a plane-world. Let us represent its inhabitants by thin pieces of paper, which are free to move over the surface of the table, but cannot rise from it. Also, let the thickness (*i.e.*, height above the surface) of these beings be so small that they cannot discern it. Lastly let us premise there is no attraction in their world, so that they have not any up and down.

Placing Cube 1 in front of us, let us now ask how a plane-being could apprehend such a cube. The Black face he could easily study. He would find it bounded by Gold point, Orange line, Fawn point, Crimson line, and so on. And he would discover it was Black by cutting through any of these lines and entering it. (This operation would be equivalent to the mining of a solid being).

But of what came above the Black square he would be completely ignorant. Let us now suppose a square hole to be made in the table, so that the cube could pass through, and let the cube fit the opening so exactly that no trace of the cutting of the table be visible to the plane-being. If the cube began to pass through, it would seem to him simply to change, for of its motion he could not be aware, as he would not know the direction in which it moved. Let it pass down till the White square be just on a level with the surface of the table. The plane-being would then perceive a Light-blue point, a Reddish line, a Dull-purple point, a Bright-blue line, and so on. These would surround a White square, which belonged to the same body as that to which the Black square belonged. But in this body there would be a dimension, which was not in the square. Our upward direction would not be apprehended by him directly. Motion from above downwards would only be apprehended as a change in the figure before him. He would not say that he had before him different sections of a cube, but only a changing square. If he wanted to look at the upper square, he could only do so when the Black square had gone an inch below his plane. To study the upper square simultaneously with the lower, he would have to make a model of it, and then he could place it beside the lower one.

Looking at the cube, we see that the Reddish line corresponds precisely to the Orange line, and the Deep-yellow to the Blue line. But if the plane-being had a model of the upper square, and placed it on the right-hand side of the Black square, the Deep-yellow line would come next to the Crimson line of the Black square. There would be a discontinuity about it. All that he could do would be to observe which part in the one square corresponded to which part in the other. Obviously too there lies something between the Black square and the White.

The plane-being would notice that when a line moves in a direction not its own, it traces out a square. When the Orange line is moved away, it traces out the Black square. The conception of a new direction thus obtained, he would understand that the Orange line moving so would trace out a square, and the Blue line moving so would do the same. To us these squares are visible as wholes, the Dark-blue, and the Vermilion. To him they would be matters of verbal definition rather than ascertained facts. However, given that he had the experience of a cube being pushed through his plane, he would know there was some figure, whereof his square was part, which was bounded by his square on one side, and by a White square on another side. We have supposed him to make models of these boundaries, a Black square and a White square. The Black square, which is his solid matter, is only one boundary of a figure in Higher Space.

But we can suppose the cube to be presented to him otherwise than by passing through his plane. It can be turned round the Orange line, in which case the Blue line goes out, and, after a time, the Brown line comes in. It must be noticed that the Brown line comes into a direction opposite to that in which the Blue line ran. These two lines are at right angles to each other, and, if one be moved upwards till it is at right angles to the surface of the table, the other comes on to the surface, but runs in a direction opposite to that in which the first ran. Thus, by turning the cube about the Orange line and the Blue line, different sides of it can be shown to a plane-being. By combining the two processes of turning and pushing through the plane, all the sides can be shown to the plane-being. For instance, if the cube be turned so that the Dark-blue square be on the plane, and it be then passed through, the Light-yellow square will come in.

Now, if the plane-being made a set of models of these different appearances and studied them, he could form some rational idea of the Higher Solid which produced them. He would become able to give some consistent account of the properties of this new kind of existence; he could say what came into his plane space, if the other space penetrated the plane edge-wise or corner-wise, and could describe all that would come in as it turned about in any way.

He would have six models. Let us consider two of them—the Black and the White squares. We can observe them on the cube. Every colour on the one is different from every colour on the other. If we now ask what lies between the Orange line and the Reddish line, we know it is a square, for the Orange line moving in any direction gives a square. And, if the six models were before the plane-being, he could easily select that which showed what he wanted. For that which lies between Orange line and Reddish line must be bounded by Orange and Reddish lines. He would search among the six models lying beside each other on his plane, till he found the Dark-blue square. It is evident that only one other square differs in all its colours from the Black square, viz., the White square. For it is entirely separate. The others meet it in one of their lines. This total difference exists in all the pairs of opposite surfaces on the cube.

Now, suppose the plane-being asked himself what would appear if the cube turned round the Blue line. The cube would begin to pass through his space. The Crimson line would disappear beneath the plane and the Blue-green square would cut it, so that opposite to the Blue line in the plane there would be a Blue-green line. The French-grey line and the Dark-slate line would be cut in points, and from the Gold point to the French-grey point would be a Dark-blue line; and opposite to it would be a Light-yellow line, from the Buff point to the Dark-slate point. Thus the figure in the plane world would be an oblong instead of a square, and the interior of it would be of the same Light-buff colour as the interior of the cube. It is assumed that the plane closes up round the passing cube, as the surface of a liquid does round any object immersed.

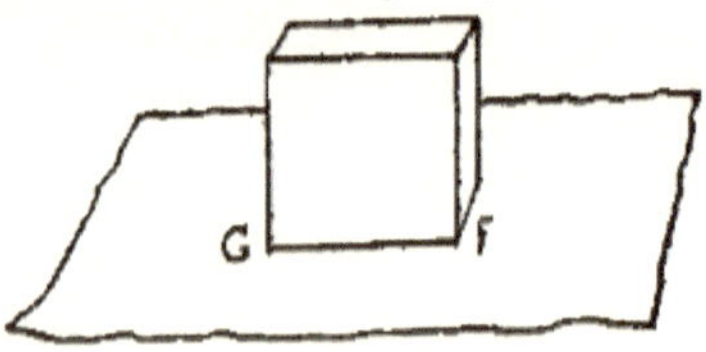


Fig.

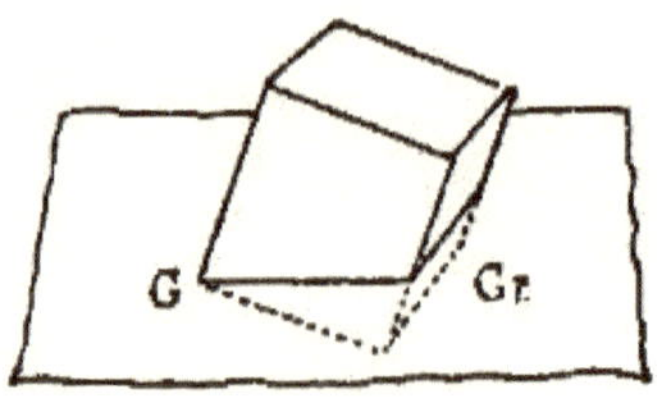


Fig. 2.

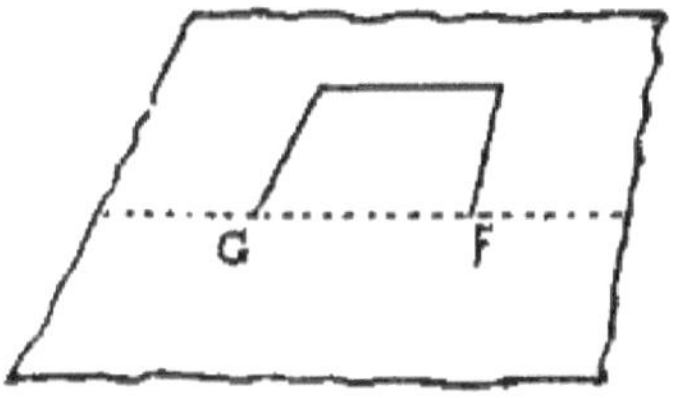


Fig. 3.

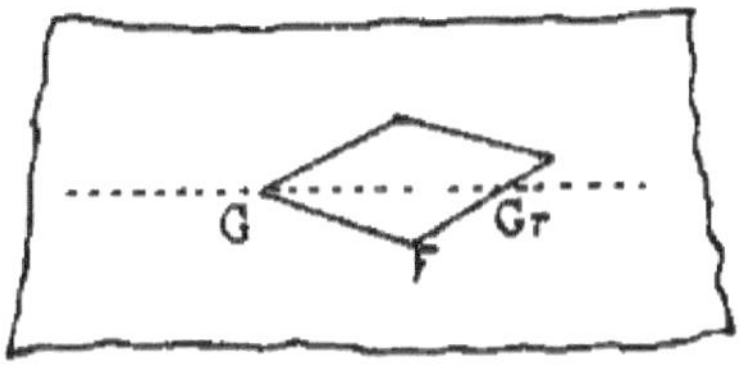


Fig. 4.

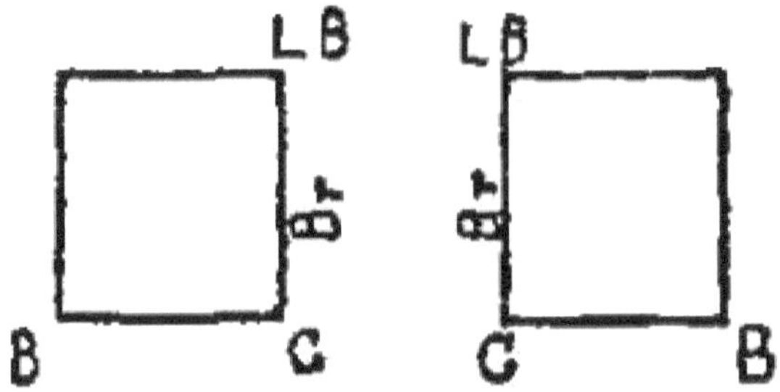


Fig. 5.

But, in order to apprehend what would take place when this twisting round the Blue line began, the plane-being would have to set to work by parts. He has no conception of what a solid would do in twisting, but he knows what a plane does. Let him, then, instead of thinking of the whole Black square, think only of the Orange line. The Dark-blue square stands on it. As far as this square is concerned, twisting round the Blue line is the same as twisting round the Gold point. Let him imagine himself in that plane at right angles to his plane-world, which contains the Dark-blue square. Let him keep his attention fixed on the line where the two planes meet, viz., that which is at first marked by the Orange line. We will call this line the line of his plane, for all that he knows of his own plane is this line. Now, let the Dark-blue square turn round the Gold point. The Orange line at once dips below the line of his plane, and the Dark-blue square passes through it. Therefore, in his plane he will see a Dark-blue line in place of the Orange one. And in place of the Fawn point, only further off from the Gold point, will be a French-grey point. The Diagrams (1), (2) show how the cube appears as it is before and after the turning. G is the Gold, F the

Fawn point. In (2) G is unmoved, and the plane is cut by the French-grey line, Gr.

Instead of imagining a direction he did not know, the plane-being could think of the Dark-blue square as lying in his plane. But in this case the Black square would be out off his plane, and only the Orange line would remain in it. Diagram (3) shows the Dark-blue square lying in his plane, and Diagram (4) shows it turning round the Gold point. Here, instead of thinking about his plane and also that at right angles to it, he has only to think how the square turning round the Gold point will cut the line, which runs left to right from G, viz., the dotted line. The French-grey line is cut by the dotted line in a point. To find out what would come in at other parts, he need only treat a number of the plane sections of the cube perpendicular to the Black square in the same manner as he had treated the Dark-blue square. Every such section would turn round a point, as the whole cube turned round the Blue line. Thus he would treat the cube as a number of squares by taking parallel sections from the Dark-blue to the Light-yellow square, and he would turn each of these round a corner of the same colour as the Blue line. Combining these series of appearances, he would discover what came into his plane as the cube turned round the Blue line. Thus, the problem of the turning of the cube could be settled by the consideration of the turnings of a number of squares.

As the cube turned, a number of different appearances would be presented to the plane-being. The Black square would change into a Light-buff oblong, with Dark-blue, Blue-green, Light-yellow, and Blue sides, and would gradually elongate itself until it became as long as the diagonal of the square side of the cube; and then the bounding line opposite to the Blue line would change from Blue-green to Bright-blue, the other lines remaining the same colour. If the cube then turned still further, the Bright-blue line would become White, and the oblong would diminish in length. It would in time become a Vermilion square, with a Deep-yellow line opposite to the Blue line. It would then pass wholly below the plane, and only the Blue line would remain.

If the turning were continued till half a revolution had been accomplished, the Black square would come in again. But now it would come up into the plane from underneath. It would appear as a Black square exactly similar to the first; but the Orange line, instead of running left to right from Gold point, would run right to left. The square would be the same, only differently disposed with regard to the Blue line. It would be the looking-glass image of the first square. There would be a difference in respect of the lie of the particles of which it was composed. If the plane-being could examine its thickness, he would find that particles which, in the first case, lay above others, now lay below them. But, if he were really a plane-being, he would have no idea of thickness in his squares, and he would find them both quite identical. Only the one would be to the other as if it had been pulled through itself. In this phenomenon of symmetry he would apprehend the difference of the lie of the line, which went in the, to him, unknown direction of up-and-down.

## CHAPTER II.
## FURTHER APPEARANCES OF A CUBE TO A PLANE-BEING.

Before leaving the observation of the cube, it is well to look at it for a moment as it would appear to a plane-being, in whose world there was such a fact as attraction. To do this, let the cube rest on the table, so that its Dark-blue face is perpendicular in front of us. Now, let a sheet of paper be placed in contact with the Dark-blue square. Let up and sideways be the two dimensions of the plane-being, and away the unknown direction. Let the line where the paper meets the table, represent the surface of his earth. Then, there is to him, as all that he can apprehend of the cube, a Dark-blue square standing upright; and, when we look over the edge of the paper, and regard merely the part in contact with the paper, we see what the plane-being would see.

If the cube be turned round the up line, the Brown line, the Orange line will pass to the near side of the paper, and the section made by the cube in the paper will be an oblong. Such an oblong can be cut out; and when the cube is fitted into it, it can be seen that it is bounded by a Brown line and a Blue-green line opposite thereto, while the other boundaries are Black and White lines. Next, if we take a section half-way between the Black and White squares, we shall have a square cutting the plane of the aforesaid paper in a single line. With regard to this section, all we have to inquire is, What will take the place of this line as the cube turns? Obviously, the line will elongate. From a Dark-blue line it will change to a Light-buff line, the colour of the inside of the section, and will terminate in a Blue-green point instead of a French-grey. Again, it is obvious that, if the cube turns round the Orange line, it will give rise to a series of oblongs, stretching upwards. This turning can be continued till the cube is wholly on the near side of the paper, and only the Orange line remains. And, when the cube has made half a revolution, the Dark-blue square will return into the plane; but it will run downwards instead of upwards as at first. Thereafter, if the cube turn further, a series of oblongs will appear, all running downwards from the Orange line. Hence, if all the appearances produced by the revolution of the cube have to be shown, it must be supposed to be raised some distance above the plane-being's earth, so that those appearances may be shown which occur when it is turned round the Orange line downwards, as well as when it is turned upwards. The unknown direction comes into the plane either upwards or downwards, but there is no special connection between it and either of these directions. If it come in upwards, the Brown line goes nearwards or -Y; if it come in downwards, or -Z, the Brown line goes away, or Y.

Let us consider more closely the directions which the plane-being would have. Firstly, he would have up-and-down, that is, away from his earth and towards it on the plane of the paper, the surface of his earth being the line where the paper meets the table. Then, if he moved along the surface of his earth, there would only be a line for him to move in, the line running right and left. But, being the direction of his movement, he would say it ran forwards and backwards. Thus he would simply have the words up and down, forwards and backwards, and the expressions right and left would have no meaning for him. If he were to frame a notion of a world in higher dimensions, he must invent new words for distinctions not within his experience.

To repeat the observations already made, let the cube be held in front of the observer, and suppose the Dark-blue square extended on every side so as to form a plane. Then let this plane be considered as independent of the Dark-blue square. Now, holding the Brown line between finger and thumb, and touching its extremities, the Gold and Light-blue points, turn the cube round the Brown line. The Dark-blue square will leave the plane, the Orange line will tend towards the -Y direction, and the Blue line will finally come into the plane pointing in the +X direction. If we move the cube so that the line which leaves the plane runs +Y, then the line which before ran +Y will come into the plane in the direction opposite to that of the line which has left the plane. The Blue line, which runs in the unknown direction can come into either of the two known directions of the plane. It can take the place of the Orange line by turning the cube round the Brown line, or the place of the Brown line by turning it round the Orange line. If the plane-being made models to represent these two appearances of the cube, he would have identically the same line, the Blue line, running in one of his known directions in the first model, and in the other of his known directions in the second. In studying the cube he would find it best to turn it so that the line of unknown direction ran in that direction in the positive sense. In that case, it would come into the plane in the negative sense of the known directions.

Starting with the cube in front of the observer, there are two ways in which the Vermilion square can be brought into the imaginary plane, that is the extension of the Dark-blue square. If the cube turn round the Brown line so that the Orange line goes away, (*i.e.* +Y), the Vermilion square comes in on the left of the Brown line. If it turn in the opposite direction, the Vermilion square comes in on the right of the Brown line. Thus, if we identify the plane-being with the Brown line, the Vermilion square would appear either behind or before him. These two appearances of the

Vermilion square would seem identical, but they could not be made to coincide by any movement in the plane. The diagram (Fig. 5.) shows the difference in them. It is obvious that no turn in the plane could put one in the place of the other, part for part. Thus the plane-being apprehends the reversal of the unknown direction by the disposition of his figures. If a figure, which lay on one side of a line, changed into an identical figure on the other side of it, he could be sure that a line of the figure, which at first ran in the positive unknown direction, now ran in the negative unknown direction.

We have dwelt at great length on the appearances, which a cube would present to a plane-being, and it will be found that all the points which would be likely to cause difficulty hereafter, have been explained in this obvious case.

There is, however, one other way, open to a plane-being of studying a cube, to which we must attend. This is, by steady motion. Let the cube come into the imaginary plane, which is the extension of the Dark-blue square, *i.e.* let it touch the piece of paper which is standing vertical on the table. Then let it travel through this plane at right angles to it at the rate of an inch a minute. The plane-being would first perceive a Dark-blue square, that is, he would see the coloured lines bounding that square, and enclosed therein would be what he would call a Dark-blue solid. In the movement of the cube, however, this Dark-blue square would not last for more than a flash of time. (The edges and points on the models are made very large; in reality they must be supposed very minute.) This Dark-blue square would be succeeded by one of the colour of the cube's interior, *i.e.* by a Light-buff square. But this colour of the interior would not be visible to the plane-being. He would go round the square on his plane, and would see the bounding lines, *viz.* Vermilion, White, Blue-green, Black. And at the corners he would see Deep-yellow, Bright-blue, Crimson, and Blue points. These lines and points would really be those parts of the faces and lines of the cube, which were on the point of passing through his plane. Now, there would be one difference between the Dark-blue square and the Light-buff with their respective boundaries. The first only lasted for a flash; the second would last for a minute or all but a minute. Consider the Vermilion square. It appears to the plane-being as a line. The Brown line also appears to him as a line. But there is a difference between them. The Brown line only lasts for a flash, whereas the Vermilion line lasts for a minute. Hence, in this mode of presentation, we may say that for a plane-being a lasting line is the mode of apprehending a plane, and a lasting plane (which is a plane-being's solid) is the mode of apprehending our solids. In the same way, the Blue line, as it passes through his plane, gives rise to a point. This point lasts for a minute, whereas the Gold point only lasted for a flash.

## CHAPTER III.
## FOUR-SPACE. GENESIS OF A TESSARACT. ITS REPRESENTATION IN THREE-SPACE.

Hitherto we have only looked at Model 1. This, with the next seven, represent what we can observe of the simplest body in Higher Space. A few words will explain their construction. A point by its motion traces a line. A line by its motion traces either a longer line or an area; if it moves at right angles to its own direction, it traces a rectangle. For the sake of simplicity, we will suppose all movements to be an inch in length and at right angles to each other. Thus, a point moving traces a line an inch long; a line moving traces a square inch; a square moving traces a cubic inch. In these cases each of these movements produces something intrinsically different from what we had before. A square is not a longer line, nor a cube a larger square. When the cube moves, we are unable to see any new direction in which it can move, and are compelled to make it move in a direction which has previously been used. Let us suppose there is an unknown direction at right angles to all our known directions, just as a third direction would be unknown to a being confined to the surface of the table. And let the cube move in this unknown direction for an inch. We call the figure it traces a Tessaract. The models are representations of the appearances a Tessaract would present to us if shown in various ways. Consider for a moment what happens to a square when moved to form a cube. Each of its lines, moved in the new direction, traces a square; the square itself traces a new figure, a cube, which ends in another square. Now, our cube, moved in a new direction, will trace a tessaract, whereof the cube itself is the beginning, and another cube the end. These two cubes are to the tessaract as the Black square and White square are to the cube. A plane-being could not see both those squares at once, but he could make models of them and let them both rest in his plane at once. So also we can make models of the beginning and end of the tessaract. Model 1 is the cube, which is its beginning; Model 2 is the cube which is its end. It will be noticed that there are no two colours alike in the two models. The Silver point corresponds to the Gold point, that is, the Silver point is the termination of the line traced by the Gold point moving in the new direction. The sides correspond in the following manner:—

SIDES.

| *Model 1.* | | *Model 2.* |
|---|---|---|
| Black | corresponds to | Bright-green |
| White | „ „ | Light-grey |
| Vermilion | „ „ | Indian-red |
| Blue-green | „ „ | Yellow-ochre |

| | | |
|---|---|---|
| Dark-blue | ,, ,, | Burnt-sienna |
| Light-yellow | ,, ,, | Dun |

The two cubes should be looked at and compared long enough to ensure that the corresponding sides can be found quickly. Then there are the following correspondencies in points and lines.

### POINTS.

| *Model 1.* | | *Model 2.* |
|---|---|---|
| Gold | corresponds to | Silver |
| Fawn | ,, ,, | Turquoise |
| Terra-cotta | ,, ,, | Earthen |
| Buff | ,, ,, | Blue tint |
| Light-blue | ,, ,, | Quaker-green |
| Dull-purple | ,, ,, | Peacock-blue |
| Deep-blue | ,, ,, | Orange-vermilion |
| Red | ,, ,, | Purple |

### LINES

| *Model 1.* | | *Model 2.* |
|---|---|---|
| Orange | corresponds to | Leaf-green |
| Crimson | ,, ,, | Dull-green |
| Green-grey | ,, ,, | Dark-purple |
| Blue | ,, ,, | Purple-brown |
| Brown | ,, ,, | Dull-blue |
| French-grey | ,, ,, | Dark-pink |
| Dark-slate | ,, ,, | Pale-pink |
| Green | ,, ,, | Indigo |
| Reddish | ,, ,, | Brown-green |
| Bright-blue | ,, ,, | Dark-green |
| Leaden | ,, ,, | Pale-yellow |
| Deep-yellow | ,, ,, | Dark |

The colour of the cube itself is invisible, as it is covered by its boundaries. We suppose it to be Sage-green.

These two cubes are just as disconnected when looked at by us as the black and white squares would be to a plane-being if placed side by side on his plane. He cannot see the squares in their right position with regard to each other, nor can we see the cubes in theirs.

Let us now consider the vermilion side of Model 1. If it move in the X direction, it traces the cube of Model 1. Its Gold point travels along the Orange line, and itself, after tracing the cube, ends in the Blue-green square. But if it moves in the new direction, it will also trace a cube, for the new direction is at right angles to the up and away directions, in which the Brown and Blue lines run. Let this square, then, move in the unknown direction, and trace a cube. This cube we cannot see, because the unknown direction runs out of our space at once, just as the up direction runs out of the plane of the table. But a plane-being could see the square, which the Blue line traces when moved upwards, by the cube being turned round the

Blue line, the Orange line going upwards; then the Brown line comes into the plane of the table in the -X direction. So also with our cube. As treated above, it runs from the Vermilion square out of our space. But if the tessaract were turned so that the line which runs from the Gold point in the unknown direction lay in our space, and the Orange line lay in the unknown direction, we could then see the cube which is formed by the movement of the Vermilion square in the new direction.

Take Model 5. There is on it a Vermilion square. Place this so that it touches the Vermilion square on Model 1. All the marks of the two squares are identical. This Cube 5, is the one traced by the Vermilion square moving in the unknown direction. In Model 5, the whole figure, the tessaract, produced by the movement of the cube in the unknown direction, is supposed to be so turned that the Orange line passes into the unknown direction, and that the line which went in the unknown direction, runs opposite to the old direction of the Orange line. Looking at this new cube, we see that there is a Stone line running to the left from the Gold point. This line is that which the Gold point traces when moving in the unknown direction.

It is obvious that, if the Tessaract turns so as to show us the side, of which Cube 5 is a model, then Cube 1 will no longer be visible. The Orange line will run in the unknown or fourth direction, and be out of our sight, together with the whole cube which the Vermilion square generates, when the Gold point moves along the Orange line. Hence, if we consider these models as real portions of the tessaract, we must not have more than one before us at once. When we look at one, the others must necessarily be beyond our sight and touch. But we may consider them simply as models, and, as such, we may let them lie alongside of each other. In this case, we must remember that their real relationships are not those in which we see them.

We now enumerate the sides of the new Cube 5, so that, when we refer to it, any colour may be recognised by name.

The square Vermilion traces a Pale-green cube, and ends in an Indian-red square.

(The colour Pale-green of this cube is not seen, as it is entirely surrounded by squares and lines of colour.)

Each Line traces a Square and ends in a Line.

| | | | | | | | | |
|---|---|---|---|---|---|---|---|---|
| The | Blue | line | traces a | Light-brown | square | and ends in a | Purple-brown | line |
| „ | Brown | „ | | Yellow | „ | | Dull-blue | „ |
| „ | Deep-yellow | „ | | Light-red | „ | | Dark | „ |
| „ | Green | „ | | Deep-crimson | „ | | Indigo | „. |

Each Point traces a Line and ends in a Point.

| | | | | | | | |
|---|---|---|---|---|---|---|---|
| The | Gold | point | traces a | Stone | line | and ends in a | Silver point |
| „ | Buff | „ | | Light-green | „ | | Blue-tint „ |
| „ | Light-blue | „ | | Rich-red | „ | | Quaker-green „ |
| „ | Red | „ | | Emerald | „ | | Purple „. |

It will be noticed that besides the Vermilion square of this cube another square of it has been seen before. A moment's comparison with the experience of a plane-being will make this more clear. If a plane-being has before him models of the Black and White squares of the Cube, he sees that all the colours of the one are different from all the colours of the other. Next, if he looks at a model of the Vermilion square, he sees that it starts from the Blue line and ends in a line of the White square, the Deep-yellow line. In this square he has two lines which he had before, the Blue line with Gold and Buff points, the Deep-yellow line with Light-blue and Red points. To him the Black and White squares are his Models 1 and 2, and the Vermilion square is to him as our Model 5 is to us. The left-hand square of Model 5 is Indian-red, and is identical with that of the same colour on the left-hand side of Model 2. In fact, Model 5 shows us what lies between the Vermilion face of 1, and the Indian-red face of 2.

From the Gold point we suppose four perfectly independent lines to spring forth, each of them at right angles to all the others. In our space there is only room for three lines mutually at right angles. It will be found, if we try to introduce a fourth at right angles to each of three, that we fail; hence, of these four lines one must go out of the space we know. The colours of these four lines are Brown, Orange, Blue, Stone. In Model 1 are shown the Brown, Orange, and Blue. In Model 5 are shown the Brown, Blue, and Stone. These lines might have had any directions at first, but we chose to begin with the Brown line going up, or Z, the Orange going X, the Blue going Y, and the Stone line going in the unknown direction, which we will call W.

Consider for a moment the Stone and the Orange lines. They can be seen together on Model 7 by looking at the lower face of it. They are at right angles to each other, and if the Orange line be turned to take the place of the Stone line, the latter will run into the negative part of the direction previously occupied by the former. This is the reason that the Models 3, 5, and 7 are made with the Stone line always running in the reverse direction of that line of Model 1, which is wanting in each respectively. It will now be easy to find out Models 3 and 7. All that has to be done is, to discover what faces they have in common with 1 and 2, and these faces will show from which planes of 1 they are generated by motion in the unknown direction.

Take Model 7. On one side of it there is a Dark-blue square, which is identical with the Dark-blue square of Model 1. Placing it so that it coincides with 1 by this square line for line, we see that the square nearest to us is Burnt-sienna, the same as the near square on Model 2. Hence this cube is a model of what the Dark-blue square traces on moving in the unknown direction. Here the unknown direction coincides with the negative away direction. In fact, to see this cube, we have been obliged to suppose the Blue line turned into the unknown direction, for we cannot look at more than three of these rectangular lines at once in our space, and in this Model 7 we have the Brown, Orange, and Stone lines. The faces, lines, and points of Cube 7 can be identified by the following list.

The Dark-blue square traces a Dark-stone cube (whose interior is rendered invisible by the bounding squares), and ends in a Burnt-sienna square.

Each Line traces a Square and ends in a Line.

| | | | | | | | | |
|---|---|---|---|---|---|---|---|---|
| The | Orange | line | | Azure | square | | Leaf-green | line |
| „ | Brown | „ | | Yellow | „ | | Dull-blue | „ |
| „ | French-grey | „ | traces an | Yellow-green | „ | and ends in a | Dark-pink | „ |
| „ | Reddish | „ | | Ochre | „ | | Brown-green | „. |

Each Point traces a Line and ends in a Point.

| | | | | | | | | |
|---|---|---|---|---|---|---|---|---|
| The | Gold | point | | Stone | line | | Silver | point |
| „ | Fawn | „ | | Smoke | „ | | Turquoise | „ |
| „ | Light-blue | „ | traces a | Rich-red | „ | and ends in a | Quaker-green | „ |
| „ | Dull-purple | „ | | Green-blue | „ | | Peacock-blue | „. |

If we now take Model 3, we see that it has a Black square uppermost, and has Blue and Orange lines. Hence, it evidently proceeds from the Black square in Model 1; and it has in it Blue and Orange lines, which proceed from the Gold point. But besides these, it has running downwards a Stone line. The line wanting is the Brown line, and, as in the other cases, when one of the three lines of Model 1 turns out into the unknown direction, the Stone line turns into the direction opposite to that from which the line has turned. Take this Model 3 and place it underneath Model 1, raising the latter so that the Black squares on the two coincide line for line. Then we see what would come into our view if the Brown line were to turn into the unknown direction, and the Stone line come into our space downwards. Looking at this cube, we see that the following parts of the tessaract have been generated.

The Black square traces a Brick-red cube (invisible because covered by its own sides and edges), and ends in a Bright-green square.

Each Line traces a Square and ends in a Line.

| | | | | | | | | |
|---|---|---|---|---|---|---|---|---|
| The | Orange | line | | Azure | square | | Leaf-green | line |
| „ | Crimson | „ | traces an | Rose | „ | and ends in a | Dull-green | „ |
| „ | Green-grey | „ | | Sea-blue | „ | | Dark-purple | „ |

| | | | | | | |
|---|---|---|---|---|---|---|
| „ Blue | „ | | Light-brown | „ | Purple-brown | „. |

Each Point traces a Line and ends in a Point.

| | | | | | | | | |
|---|---|---|---|---|---|---|---|---|
| The Gold | point | traces a | Stone | line | and ends in a | Silver | point |
| „ Fawn | „ | | Smoke | „ | | Turquoise | „ |
| „ Terra-cotta | „ | | Magenta | „ | | Earthen | „ |
| „ Buff | „ | | Light-green | „ | | Blue-tint | „. |

This completes the enumeration of the regions of Cube 3. It may seem a little unnatural that it should come in downwards; but it must be remembered that the new fourth direction has no more relation to up-and-down than to right-and-left or to near-and-far.

And if, instead of thinking of a plane-being as living on the surface of a table, we suppose his world to be the surface of the sheet of paper touching the Dark-blue square of Cube 1, then we see that a turn round the Orange line, which makes the Brown line go into the plane-being's unknown direction, brings the Blue line into his downwards direction.

There still remain to be described Models 4, 6, and 8. It will be shown that Model 4 is to Model 3 what Model 2 is to Model 1. That is, if, when 3 is in our space, it be moved so as to trace a tessaract, 4 will be the opposite cube in which the tessaract ends. There is no colour common to 3 and 4. Similarly, 6 is the opposite boundary of the tessaract generated by 5, and 8 of that by 7.

A little closer consideration will reveal several points. Looking at Cube 5, we see proceeding from the Gold point a Brown, a Blue, and a Stone line. The Orange line is wanting; therefore, it goes in the unknown direction. If we want to discover what exists in the unknown direction from Cube 5, we can get help from Cube 1. For, since the Orange line lies in the unknown direction from Cube 5, the Gold point will, if moved along the Orange line, pass in the unknown direction. So also, the Vermilion square, if moved along in the direction of the Orange line, will proceed in the unknown direction. Looking at Cube 1 we see that the Vermilion square thus moved ends in a Blue-green square. Then, looking at Model 6, on it, corresponding to the Vermilion square on Cube 5, is a Blue-green square.

Cube 6 thus shows what exists an inch beyond 5 in the unknown direction. Between the right-hand face on 5 and the right-hand face on 6 lies a cube, the one which is shown in Model 1. Model 1 is traced by the Vermilion square moving an inch along the direction of the Orange line. In Model 5, the Orange line goes in the unknown direction; and looking at Model 6 we see what we should get at the end of a movement of one inch in that direction. We have still to enumerate the colours of Cubes 4, 6, and 8, and we do so in the following list. In the first column is designated the part of the cube; in the columns under 4, 6, 8, come the colours which 4, 6,

8, respectively have in the parts designated in the corresponding line in the first column.

Cube itself:—

| | 4 | 6 | 8 |
|---|---|---|---|
| | Chocolate | Oak-yellow | Salmon |

Squares:—

| | 4 | 6 | 8 |
|---|---|---|---|
| Lower face | Light-grey | Rose | Sea-blue |
| Upper | White | Deep-brown | Deep-green |
| Left-hand | Light-red | Yellow-ochre | Deep-crimson |
| Right-hand | Deep-brown | Blue-green | Dark-grey |
| Near | Ochre | Yellow-green | Dun |
| Far | Deep-green | Dark-grey | Light-yellow |

Lines:—

On ground, going round the square from left to right:—

| | 4 | 6 | 8 |
|---|---|---|---|
| 1. | Brown-green | Smoke | Dark-purple |
| 2. | Dark-green | Crimson | Magenta |
| 3. | Pale-yellow | Magenta | Green-grey |
| 4. | Dark | Dull-green | Light-green |

Vertical, going round the sides from left to right:—

| | 4 | 6 | 8 |
|---|---|---|---|
| 1. | Rich-red | Dark-pink | Indigo |
| 2. | Green-blue | French-grey | Pale-pink |
| 3. | Sea-green | Dark-slate | Dark-slate |
| 4. | Emerald | Pale-pink | Green |

Round upper face in same order:—

| | 4 | 6 | 8 |
|---|---|---|---|
| 1. | Reddish | Green-blue | Pale-yellow |
| 2. | Bright-blue | Bright-blue | Sea-green |
| 3. | Leaden | Sea-green | Leaden |
| 4. | Deep-yellow | Dark-green | Emerald |

Points:—

On lower face, going from left to right:—

| | 4 | 6 | 8 |
|---|---|---|---|
| 1. | Quaker-green | Turquoise | Blue-tint |
| 2. | Peacock-blue | Fawn | Earthen |
| 3. | Orange-vermilion | Terra-cotta | Terra-cotta |
| 4. | Purple | Earthen | Buff |

On upper face:—

| | 4 | 6 | 8 |
|---|---|---|---|
| 1. | Light-blue | Peacock-blue | Purple |
| 2. | Dull-purple | Dull-purple | Orange-vermilion |
| 3. | Deep-blue | Deep-blue | Deep-blue |
| 4. | Red | Orange-vermilion | Red |

If any one of these cubes be taken at random, it is easy enough to find out to what part of the Tessaract it belongs. In all of them, except 2, there will be one face, which is a copy of a face on 1; this face is, in fact, identical with the face on 1 which it resembles. And the model shows what lies in the unknown direction from that face. This unknown direction is turned into our space, so that we can see and touch the result of moving a square in it. And we have sacrificed one of the three original directions in order to do this. It will be found that the line, which in 1 goes in the 4th direction, in the other models always runs in a negative direction.

Let us take Model 8, for instance. Searching it for a face we know, we come to a Light-yellow face away from us. We place this face parallel with the Light-yellow face on Cube 1, and we see that it has a Green line going up, and a Green-grey line going to the right from the Buff point. In these respects it is identical with the Light-yellow face on Cube 1. But instead of a Blue line coming towards us from the Buff point, there is a Light-green line. This Light-green line, then, is that which proceeds in the unknown direction from the Buff point. The line is turned towards us in this Model 8 in the negative Y direction; and looking at the model, we see exactly what is formed when in the motion of the whole cube in the unknown direction, the Light-yellow face is moved an inch in that direction. It traces out a Salmon cube (*v.* Table on p. 127), and it has Sea-blue and Deep-green sides below and above, and Deep-crimson and Dark-grey sides left and right, and Dun and Light-yellow sides near and far. If we want to verify the correctness of any of these details, we must turn to Models 1 and 2. What lies an inch from the Light-yellow square in the unknown direction? Model 2 tells us, a Dun square. Now, looking at 8, we see that towards us lies a Dun square. This is what lies an inch in the unknown direction from the Light-yellow square. It is here turned to face us, and we can see what lies between it and the Light-yellow square.

## CHAPTER IV.
## TESSARACT MOVING THROUGH THREE-SPACE. MODELS OF THE SECTIONS.

In order to obtain a clear conception of the higher solid, a certain amount of familiarity with the facts shown in these models is necessary. But the best way of obtaining a systematic knowledge is shown hereafter. What these models enable us to do, is to take a general review of the subject. In all of them we see simply the boundaries of the tessaract in our space; we can no more see or touch the tessaract's solidity than a plane-being can touch the cube's solidity.

There remain the four models 9, 10, 11, 12. Model 9 represents what lies between 1 and 2. If 1 be moved an inch in the unknown direction, it traces out the tessaract and ends in 2. But, obviously, between 1 and 2 there must be an infinite number of exactly similar solid sections; these are all like Model 9.

Take the case of a plane-being on the table. He sees the Black square,—that is, he sees the lines round it,—and he knows that, if it moves an inch in some mysterious direction, it traces a new kind of figure, the opposite boundary whereof is the White square. If, then, he has models of the White and Black squares, he has before him the end and beginning of our cube. But between these squares are any number of others, the plane sections of the cube. We can see what they are. The interior of each is a Light-buff (the colour of the substance of the cube), the sides are of the colours of the vertical faces of the cube, and the points of the colours of the vertical lines of the cube, viz., Dark-blue, Blue-green, Light-yellow, Vermilion lines, and Brown, French-grey, Dark-slate, Green points. Thus, the square, in moving in the unknown direction, traces out a succession of squares, the assemblage of which makes the cube in layers. So also the cube, moving in the unknown direction, will at any point of its motion, still be a cube; and the assemblage of cubes thus placed constitutes the tessaract in layers. We suppose the cube to change its colour directly it begins to move. Its colour between 1 and 2 we can easily determine by finding what colours its different parts assume, as they move in the unknown direction. The Gold point immediately begins to trace a Stone-line. We will look at Cube 5 to see what the Vermilion face becomes; we know the interior of that cube is Pale-green (*v.* Table, p. 122). Hence, as it moves in the unknown direction, the Vermilion square forms in its course a series of Pale-green squares. The Brown line gives rise to a Yellow square; hence, at every point of its course in the fourth direction, it is a Yellow line, until, on taking its final position, it becomes a Dull-blue line. Looking at Cube 5, we see that the Deep yellow line becomes a Light-red line, the Green line a Deep Crimson one, the

Gold point a Stone one, the Light-blue point a Rich-red one, the Red point an Emerald one, and the Buff point a Light-green one. Now, take the Model 9. Looking at the left side of it, we see exactly that into which the Vermilion square is transformed, as it moves in the unknown direction. The left side is an exact copy of a section of Cube 5, parallel to the Vermilion face.

But we have only accounted for one side of our Model 9. There are five other sides. Take the near side corresponding to the Dark-blue square on Cube 1. When the Dark-blue square moves, it traces a Dark-stone cube, of which we have a copy in Cube 7. Looking at 7 (*v.* Table, p. 124), we see that, as soon as the Dark-blue square begins to move, it becomes of a Dark-stone colour, and has Yellow, Ochre, Yellow-green, and Azure sides, and Stone, Rich-red, Green-blue, Smoke lines running in the unknown direction from it. Now, the side of Model 9, which faces us, has these colours the squares being seen as lines, and the lines as points. Hence Model 9 is a copy of what the cube becomes, so far as the Vermilion and Dark-blue sides are concerned, when, moving in the unknown direction, it traces the tessaract.

We will now look at the lower square of our model. It is a Brick-red square, with Azure, Rose, Sea-blue, and Light-brown lines, and with Stone, Smoke, Magenta, and Light-green points. This, then, is what the Black square should change into, as it moves in the unknown direction. Let us look at Model 3. Here the Stone line, which is the line in the unknown direction, runs downwards. It is turned into the downwards direction, so that the cube traced by the Black square may be in our space. The colour of this cube is Brick-red; the Orange line has traced an Azure, the Blue line a Light-brown, the Crimson line a Rose, and the Green-grey line a Sea-blue square. Hence, the lower square of Model 9 shows what the Black square becomes, as it traces the tessaract; or, in other words, the section of Model 3 between the Black and Bright-green squares exactly corresponds to the lower face of Model 9.

Therefore, it appears that Model 9 is a model of a section of the tessaract, that it is to the tessaract what a square between the Black and White squares is to the cube.

To prove the other sides correct, we have to see what the White, Blue-green, and Light-yellow squares of Cube 1 become, as the cube moves in the unknown direction. This can be effected by means of the Models 4, 6, 8. Each cube can be used as an index for showing the changes through which any side of the first model passes, as it moves in the unknown direction till it becomes Cube 2. Thus, what becomes of the White square?

Look at Cube 4. From the Light-blue corner of its White square runs downwards the Rich-red line in the unknown direction. If we take a parallel section below the White square, we have a square bounded by Ochre, Deep-brown, Deep-green, and Light-red lines; and by Rich-red, Green-blue, Sea-green, and Emerald points. The colour of the cube is Chocolate, and therefore its section is Chocolate. This description is exactly true of the upper surface of Model 9.

There still remain two sides, those corresponding to the Light-yellow and Blue-green of Cube 1. What the Blue-green square becomes midway between Cubes 1 and 2 can be seen on Model 6. The colour of the last-named is Oak-yellow, and a section parallel to its Blue-green side is surrounded by Yellow-green, Deep-brown, Dark-grey and Rose lines and by Green-blue, Smoke, Magenta, and Sea-green points. This is exactly similar to the right side of Model 9. Lastly, that which becomes of the Light-yellow side can be seen on Model 8. The section of the cube is a Salmon square bounded by Deep-crimson, Deep-green, Dark-grey and Sea-blue lines and by Emerald, Sea-green, Magenta, and Light-green points.

Thus the models can be used to answer any question about sections. For we have simply to take, instead of the whole cube, a plane, and the relation of the whole tessaract to that plane can be told by looking at the model, which, starting with that plane, stretches from it in the unknown direction.

We have not as yet settled the colour of the interior of Model 9. It is that part of the tessaract which is traced out by the interior of Cube 1. The unknown direction starts equally and simultaneously from every point of every part of Cube 1, just as the up direction starts equally and simultaneously from every point of a square. Let us suppose that the cube, which is Light-buff, changes to a Wood-colour directly it begins to trace the tessaract. Then the internal part of the section between 1 and 2 will be a Wood-colour. The sides of the Model 9 are of the greatest importance. They are the colour of the six cubes, 3, 4, 5, 6, 7, and 8. The colours of 1 and 2 are wanting, viz. Light-buff and Sage-green. Thus the section between 1 and 2 can be found by its wanting the colours of the Cubes 1 and 2.

Looking at Models 10, 11, and 12 in a similar manner, the reader will find they represent the sections between Cubes 3 and 4, Cubes 5 and 6, and Cubes 7 and 8 respectively.

# CHAPTER V.
# REPRESENTATION OF THREE-SPACE BY NAMES, AND IN A PLANE.

We may now ask ourselves the best way of passing on to a clear comprehension of the facts of higher space. Something can be effected by looking at these models; but it is improbable that more than a slight sense of analogy will be obtained thus. Indeed, we have been trusting hitherto to a method which has something vicious about it—we have been trusting to our sense of what *must* be. The plan adopted, as the serious effort towards the comprehension of this subject, is to learn a small portion of higher space. If any reader feel a difficulty in the foregoing chapters, or if the subject is to be taught to young minds, it is far better to abandon all attempt to see what higher space *must* be, and to learn what it *is* from the following chapters.

## NAMING A PIECE OF SPACE.

The diagram (Fig. 6) represents a block of 27 cubes, which form Set 1 of the 81 cubes. The cubes are coloured, and it will be seen that the colours are arranged after the pattern of Model 1 of previous chapters, which will serve as a key to the block. In the diagram, G. denotes Gold, O. Orange, F. Fawn, Br. Brown, and so on. We will give names to the cubes of this block. They should not be learnt, but kept for reference. We will write these names in three sets, the lowest consisting of the cubes which touch the table, the next of those immediately above them, and the third of those at the top. Thus the Gold cube is called Corvus, the Orange, Cuspis, the Fawn, Nugæ, and the central one below, Syce. The corresponding colours of the following set can easily be traced.

| | | |
|---|---|---|
| Olus | Semita | Lama |
| Via | Mel | Iter |
| Ilex | Callis | Sors |
| | | |
| Bucina | Murex | Daps |
| Alvus | Mala | Proes |
| Arctos | Mœna | Far |
| | | |
| Cista | Cadus | Crus |
| Dos | Syce | Bolus |
| Corvus | Cuspis | Nugæ |

Thus the central or Light-buff cube is called Mala; the middle one of the lower face is Syce; of the upper face Mel; of the right face, Proes; of the left, Alvus; of the front, Mœna (the Dark-blue square of Model 1); and of the back, Murex (the Light-yellow square).

Now, if Model 1 be taken, and considered as representing a block of 64 cubes, the Gold corner as one cube, the Orange line as two cubes, the Fawn point as one cube, the Dark-blue square as four cubes, the Light-buff interior as eight cubes, and so on, it will correspond to the diagram (Fig. 7). This block differs from the last in the number of cubes, but the arrangement of the colours is the same. The following table gives the names which we will use for these cubes. There are no new names; they are only applied more than once to all cubes of the same colour.

Fourt<br>h Floor.

| Olus | Semit<br>a | Semit<br>a | Lama |
|---|---|---|---|
| Via | Mel | Mel | Iter |
| Via | Mel | Mel | Iter |
| Ilex | Callis | Callis | Sors |

Third Floor.

| Bucin<br>a | Mure<br>x | Mure<br>x | Daps |
|---|---|---|---|
| Alvus | Mala | Mala | Proes |
| Alvus | Mala | Mala | Proes |
| Arcto<br>s | Mœn<br>a | Mœn<br>a | Far |

Secon<br>d Floor.

| Bucin<br>a | Mure<br>x | Mure<br>x | Daps |
|---|---|---|---|
| Alvus | Mala | Mala | Proes |
| Alvus | Mala | Mala | Proes |
| Arcto<br>s | Mœn<br>a | Mœn<br>a | Far |

First Floor.

| Cista | Cadus | Cadus | Crus |
|---|---|---|---|
| Dos | Syce | Syce | Bolu<br>s |
| Dos | Syce | Syce | Bolu<br>s |
| Corvu<br>s | Cuspi<br>s | Cuspi<br>s | Nug<br>æ |

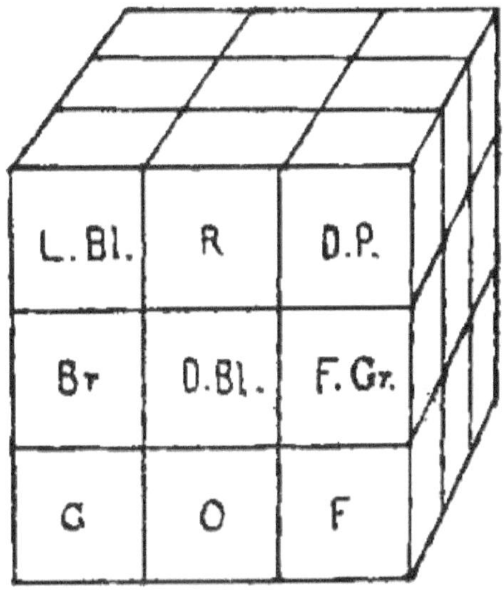
L. Bl.
R
D.P.
Br
D.Bl.
F.Gr.
G
O
F

Fig. 6

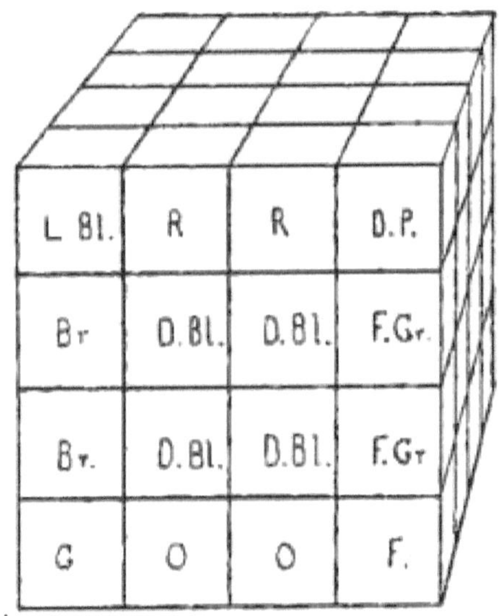
L Bl.
R
R
D.P.
Br
D.Bl.
D.Bl.
F.Gr.
Br.
D.Bl.
D.Bl.
F.Gr
G
O
O
F.

Fig. 7.

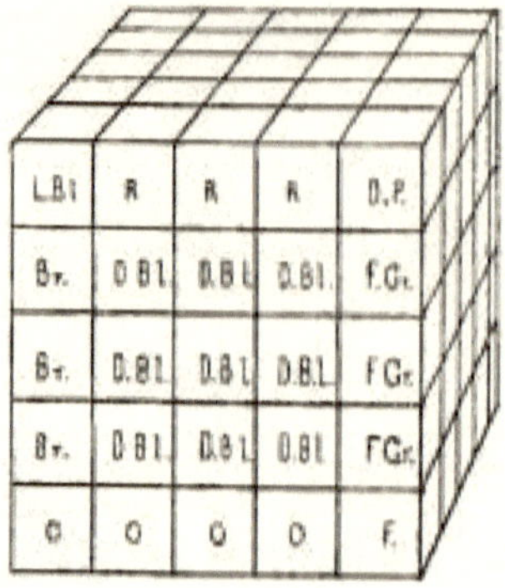


Fig. 8.

If we now consider Model 1 to represent a block, five cubes each way, built up of inch cubes, and colour it in the same way, that is, with similar colours for the corner-cubes, edge-cubes, face-cubes, and interior-cubes, we obtain what is represented in the diagram (Fig. 8). Here we have nine Dark-blue cubes called Mœna; that is, Mœna denotes the nine Dark-blue cubes, forming a layer on the front of the cube, and filling up the whole front except the edges and points. Cuspis denotes three Orange, Dos three Blue, and Arctos three Brown cubes.

Now, the block of cubes can be similarly increased to any size we please. The corners will always consist of single cubes; that is, Corvus will remain a single cubic inch, even though the block be a hundred inches each way. Cuspis, in that case, will be 98 inches long, and consist of a row of 98 cubes; Arctos, also, will be a long thin line of cubes standing up. Mœna will be a thin layer of cubes almost covering the whole front of the block; the number of them will be 98 times 98. Syce will be a similar square layer of cubes on the ground, so also Mel, Alvus, Proes, and Murex in their respective places. Mala, the interior of the cube, will consist of 98 times 98 times 98 inch cubes.

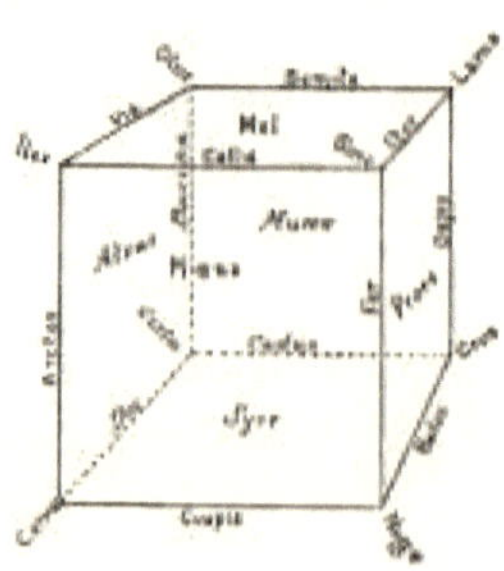

Fig. 9

Now, if we continued in this manner till we had a very large block of thousands of cubes in each side Corvus would, in comparison to the whole block, be a minute point of a cubic shape, and Cuspis would be a mere line of minute cubes, which would have length, but very small depth or height. Next, if we suppose this much sub-divided block to be reduced in size till it becomes one measuring an inch each way, the cubes of which it consists must each of them become extremely minute, and the corner cubes and line cubes would be scarcely discernible. But the cubes on the faces would be just as visible as before. For instance, the cubes composing Mœna would stretch out on the face of the cube so as to fill it up. They would form a layer of extreme thinness, but would cover the face of the cube (all of it except the minute lines and points). Thus we may use the words Corvus and Nugæ, etc., to denote the corner-points of the cube, the words Mœna, Syce, Mel, Alvus, Proes, Murex, to denote the faces. It must be remembered that these faces have a thickness, but it is extremely minute compared with the cube. Mala would denote all the cubes of the interior except those, which compose the faces, edges, and points. Thus, Mala would practically mean the whole cube except the colouring on it. And it is in this sense that these words will be used. In the models, the Gold point is intended to be a Corvus, only it is made large to be visible; so too the Orange line is meant for Cuspis, but magnified for the same reason. Finally, the 27 names of cubes, with which we began, come to be the names of the points, lines, and faces of a cube, as shown in the diagram (Fig. 9). With these names it is easy to express what a plane-being would see of any cube. Let us suppose that Mœna is only of the thickness of his matter. We suppose his matter to be composed of particles, which slip about on his plane, and are so thin that he cannot by any means discern any thickness in them. So he has no idea of thickness. But we know that his matter must have some thickness, and we suppose Mœna to be of that degree of thickness. If the cube be placed so that Mœna is in his plane, Corvus, Cuspis, Nugæ, Far, Sors, Callis, Ilex and Arctos will just come into his apprehension; they will be like bits of his matter, while all that is beyond them in the direction he does not know, will be hidden from him. Thus a plane-being can only perceive the Mœna or Syce or some one other face of a cube; that is, he would take the Mœna of a cube to be a solid in his plane-space, and he would see the lines Cuspis, Far, Callis, Arctos. To him they would bound it. The points Corvus, Nugæ, Sors, and Ilex, he would not see, for they are only as long as the thickness of his matter, and that is so slight as to be indiscernible to him.

We must now go with great care through the exact processes by which a plane-being would study a cube. For this purpose we use square slabs which have a certain thickness, but are supposed to be as thin as a plane-being's matter. Now, let us take the first set of 81 cubes again, and build them from 1 to 27. We must realize clearly that two kinds of blocks can be built. It may be built of 27 cubes, each similar to Model 1, in which case each cube has its regions coloured, but all the cubes are alike. Or it may be built of 27 differently coloured cubes like Set 1, in which case each cube is coloured wholly with one colour in all its regions. If the latter set be used, we can still use the names Mœna, Alvus, etc. to denote the front, side, etc., of any one of the cubes, whatever be its colour. When they are built up, place a piece of card against the front to represent the plane on which the plane-being lives. The front of each of the cubes in the front of the block touches the plane. In previous chapters we have supposed Mœna to be a Blue square. But we can apply the name to the front of a cube of any colour. Let us say the Mœna of each front cube is in the plane; the Mœna of the Gold cube is Gold, and so on. To represent this, take nine slabs of the same colours as the cubes. Place a stiff piece of cardboard (or a book-cover) slanting from you, and put the slabs on it. They can be supported on the incline so as to prevent their slipping down away from you by a thin book, or another sheet of cardboard, which stands for the surface of the plane-being's earth.

We will now give names to the cubes of Block 1 of the 81 Set. We call each one Mala, to denote that it is a cube. They are written in the following list in floors or layers, and are supposed to run backwards or away from the reader. Thus, in the first layer, Frenum Mala is behind or farther away than Urna Mala; in the second layer, Ostrum is in front, Uncus behind it, and Ala behind Uncus.

Third, or Top Floor.

| | | |
|---|---|---|
| Mars Mala | Merces Mala | Tyro Mala |
| Spicula Mala | Mora Mala | Oliva Mala |
| Comes Mala | Tibicen Mala | Vestis Mala |

Second, or Middle Floor.

| | | |
|---|---|---|
| Ala Mala | Cortis Mala | Aer Mala |
| Uncus Mala | Pallor Mala | Tergum Mala |
| Ostrum Mala | Bidens Mala | Scena Mala |

First, or Bottom Floor.

| | | |
|---|---|---|
| Sector Mala | Hama Mala | Remus Mala |
| Frenum Mala | Plebs Mala | Sypho Mala |
| Urna Mala | Moles Mala | Saltus Mala |

These names should be learnt so that the different cubes in the block can be referred to quite easily and immediately by name. They must be learnt in every order, that is, in each of the three directions backwards and forwards, *e.g.* Urna to Saltus, Urna to Sector, Urna to Comes; and the same reversed, viz., Comes to Urna, Sector to Urna, etc. Only by so learning them can the mind identify any one individually without even a momentary reference to the others around it. It is well to make it a rule not to proceed from one cube to a distant one without naming the intermediate cubes. For, in Space we cannot pass from one part to another without going through the intermediate portions. And, in thinking of Space, it is well to accustom our minds to the same limitations.

Urna Mala is supposed to be solid Gold an inch each way; so too all the cubes are supposed to be entirely of the colour which they show on their faces. Thus any section of Moles Mala will be Orange, of Plebs Mala Black, and so on.

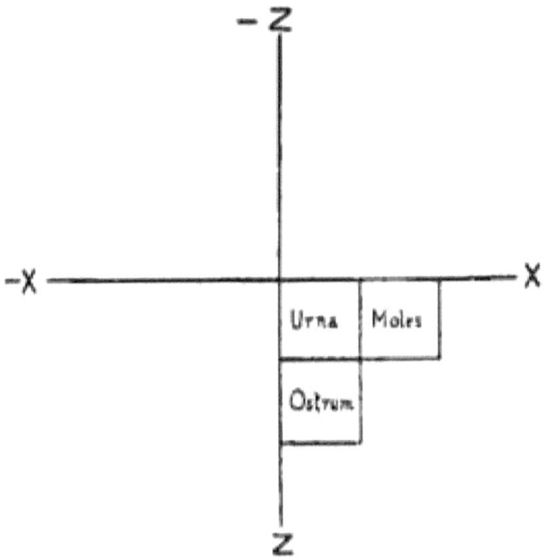


Fig. 10.

Let us now draw a pair of lines on a piece of paper or cardboard like those in the diagram (Fig. 10). In this diagram the top of the page is supposed to rest on the table, and the bottom of the page to be raised and brought near the eye, so that the plane of the diagram slopes upwards to the reader. Let Z denote the upward direction, and X the direction from left to right. Let us turn the Block of cubes with its front upon this slope *i.e.* so that Urna fits upon the square marked Urna. Moles will be to the right and Ostrum above Urna, *i.e.* nearer the eye. We might leave the block as it stands and put the piece of cardboard against it; in this case our plane-world would be vertical. It is difficult to fix the cubes in this position on the plane, and therefore more convenient if the cardboard be so inclined that they will not slip off. But the upward direction must be identified with Z. Now, taking the slabs, let us compose what a plane-being would see of the Block. He would perceive just the front faces of the cubes of the Block, as it comes into his plane; these front faces we may call the Moenas of the cubes. Let each of the slabs represent the Moena of its corresponding cube, the Gold slab of the Gold cube and so on. They are thicker than they should be; but we must overlook this and suppose we simply see the thickness as a line. We thus build a square of nine slabs to represent the appearance to a plane-being of the front face of the Block. The middle one, Bidens Moena, would be completely hidden from him by the others on all its sides, and he would see the edges of the eight outer squares. If the Block now begin to move through the plane, that is, to cut through the piece of paper at right angles to it, it will not for some time appear any different. For the sections of Urna are all Gold like the front face Moena, so that the appearance of Urna at any point in its passage will be a Gold square exactly like Urna Moena, seen by the plane-being as a line. Thus, if the speed of the Block's passage be one inch a minute, the plane-being will see no change for a minute. In other words, this set of slabs lasting one minute will represent what he sees.

When the Block has passed one inch, a different set of cubes appears. Remove the front layer of cubes. There will now be in contact with the paper nine new cubes, whose names we write in the order in which we should see them through a piece of glass standing upright in front of the Block:

| | | |
|---|---|---|
| Spicula Mala | Mora Mala | Oliva Mala |
| Uncus Mala | Pallor Mala | Tergum Mala |
| Frenum Mala | Plebs Mala | Sypho Mala |

We pick out nine slabs to represent the Moenas of these cubes, and placed in order they show what the plane-being sees of the second set of cubes as they pass through. Similarly the third wall of the Block will come into the plane, and looking at them similarly, as it were through an upright piece of glass, we write their names:

| | | |
|---|---|---|
| Mars Mala | Merces Mala | Tyro Mala |
| Ala Mala | Cortis Mala | Aer Mala |
| Sector Mala | Hama Mala | Remus Mala |

Now, it is evident that these slabs stand at different times for different parts of the cubes. We can imagine them to stand for the Moena of each cube as it passes through. In that case, the first set of slabs, which we put up, represents the Moenas of the front wall of cubes; the next set, the Moenas of the second wall. Thus, if all the three sets of slabs be together on the table, we have a representation of the sections of the cube. For some purposes it would be better to have four sets of slabs, the fourth set representing the Murex of the third wall; for the three sets only show the front faces of the cubes, and therefore would not indicate anything about the back faces of the Block. For instance, if a line passed through the Block diagonally from the point Corvus (Gold) to the point Lama (Deep-blue), it would be represented on the slabs by a point at the bottom left-hand corner of the Gold slab, a second point at the same corner of the Light-buff slab, and a third at the same corner of the Deep-blue slab. Thus, we should have the points mapped at which the line entered the fronts of the walls of cubes, but not the point in Lama at which it would leave the Block.

Let the Diagrams 1, 2, 3 (Fig. 11), be the three sets of slabs. To see the diagrams properly, the reader must set the top of the page on the table, and look along the page from the bottom of it. The line in question, which runs from the bottom left-hand near corner to the top right-hand far corner of the Block will be represented in the three sets of slabs by the points A, B, C. To complete the diagram of its course, we need a fourth set of slabs for the Murex of the third wall; the same object might be attained, if we had another Block of 27 cubes behind the first Block and represented its front or Moenas by a set of slabs. For the point, at which the line leaves the first Block is identical with that at which it enters the second Block.

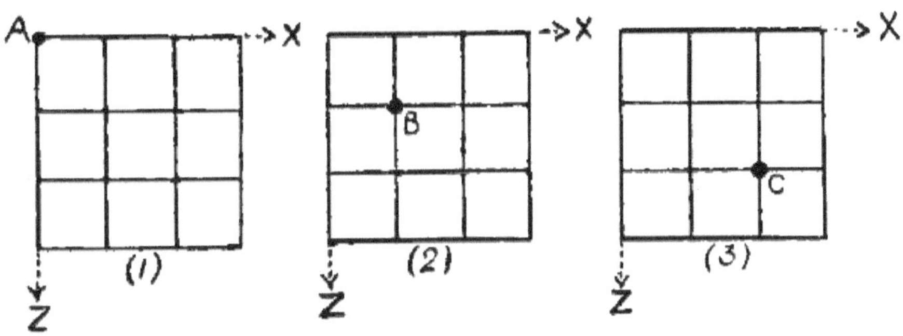


Fig. 11.

If we suppose a sheet of glass to be the plane-world, the Diagrams 1, 2, 3 (Fig. 11), may be drawn more naturally to us as Diagrams α, β, γ (Fig. 12). Here α represents the Moenas of the first wall, β those of the second, γ those of the third. But to get the plane-being's view we must look over the edge of the glass down the Z axis.

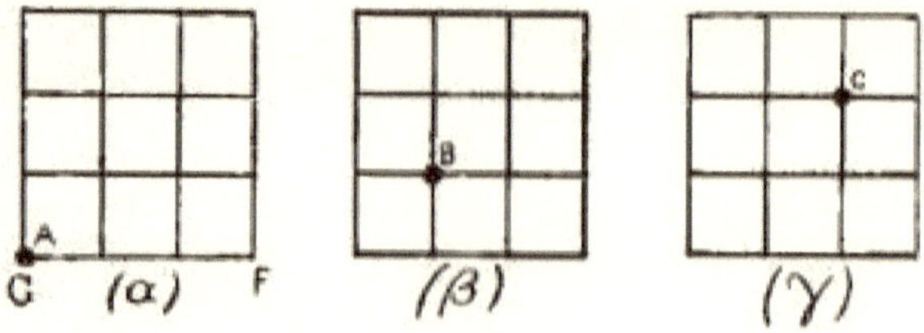


Fig. 12.

Set 2 of slabs represent the Moenas of Wall 2. These Moenas are in contact with the Murex of Wall 1. Thus Set 2 will show where the line issues from Wall 1 as well as where it enters Wall 2.

The plane-being, therefore, could get an idea of the Block of cubes by learning these slabs. He ought not to call the Gold slab Urna Mala, but Urna Moena, and so on, because all that he learns are Moenas, merely the thin faces of the cubes. By introducing the course of time, he can represent the Block more nearly. For, if he supposes it to be passing an inch a minute, he may give the name Urna Mala to the Gold slab enduring for a minute.

But, when he has learnt the slabs in this position and sequence, he has only a very partial view of the Block. Let the Block turn round the Z axis, as Model 1 turns round the Brown line. A different set of cubes comes into his plane, and now they come in on the Alvus faces. (Alvus is here used to denote the left-hand faces of the cubes, and is not supposed to be Vermilion; it is simply the thinnest slice on the left hand of the cube and of the same colour as the cube.) To represent this, the plane-being should employ a fresh set of slabs, for there is nothing common to the Moena and Alvus faces except an edge. But, since each cube is of the same colour throughout, the same slab may be used for its different faces. Thus the Alvus of Urna Mala can be represented by a Gold slab. Only it must never be forgotten that it is meant to be a new slab, and is not identical with the same slab used for Moena.

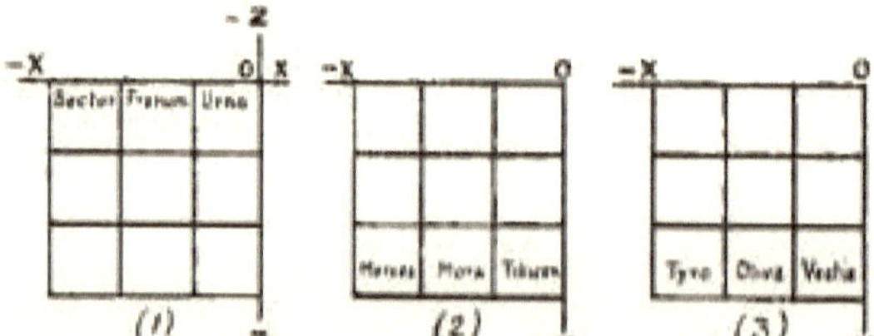


Fig. 13.

Now, when the Block of cubes has turned round the Brown line into the plane, it is clear that they will be on the side of the Z axis opposite to that on which were the Moena slabs. The line, which ran Y, now runs -X. Thus the slabs will occupy the second quadrant marked by the axes, as shown in the diagram (Fig. 13). Each of these slabs we will name Alvus. In this view, as before, the book is supposed to be tilted up towards the reader, so that the Z axis runs from O to his eye. Then, if the Block be passed at right angles through the plane, there will come into view the two sets of slabs represented in the Diagrams (Fig. 13). In copying this arrangement with the slabs, the cardboard on which they are arranged must slant upwards to the eye, *i.e.*, OZ must run up to the eye, and the sides of the slabs seen are in Diagram 2 (Fig. 13), the upper edges of Tibicen, Mora, Merces; in Diagram 3, the upper edges of Vestis, Oliva, Tyro.

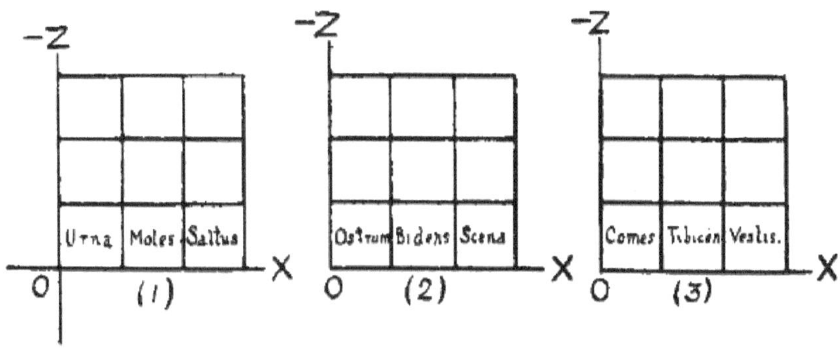


Fig. 14.

There is another view of the Block possible to a plane-being. If the Block be turned round the X axis, the lower face comes into the vertical plane. This corresponds to turning Model 1 round the Orange line. On referring to the diagram (Fig. 14), we now see that the name of the faces of the cubes coming into the plane is Syce. Here the plane-being looks from the extremity of the Z axis and the squares, which he sees run from him in the -Z direction. (As this turn of the Block brings its Syce into the vertical plane so that it extends three inches below the base line of its Moena, it is evident that the turn is only possible if the Moena be originally at a height of at least three inches above the plane-being's earth line in the vertical plane.) Next, if the Block be passed through the plane, the sections shown in the Diagrams 2 and 3 (Fig. 14) are brought into view.

Thus, there are three distinct ways of regarding the cubic Block, each of them equally primary; and if the plane-being is to have a correct idea of the Block, he must be equally familiar with each view. By means of the slabs each aspect can be represented; but we must remember in each of the three cases, that the slabs represent different parts of the cube.

When we look at the cube Pallor Mala in space, we see that it touches six other cubes by its six faces. But the plane-being could only arrive at this fact by comparing different views. Taking the three Moena sections of the Block, he can see that Pallor Mala Moena touches Plebs Moena, Mora Moena, Uncus Moena, and Tergum Moena by lines. And it takes the place of Bidens Moena, and is itself displaced by Cortis Moena as the Block passes through the plane. Next, this same Pallor Mala can appear to him as an Alvus. In this case, it touches Plebs Alvus, Mora Alvus, Bidens Alvus, and Cortis Alvus by lines, takes the place of Uncus Alvus, and is itself displaced by Tergum Alvus as the Block moves. Similarly he can observe the relations, if the Syce of the Block be in his plane.

Hence, this unknown body Pallor Mala appears to him now as one plane-figure now as another, and comes before him in different connections. Pallor Mala is that which satisfies all these relations. Each of them he can in turn present to sense; but the total conception of Pallor Mala itself can only, if at all, grow up in his mind. The way for him to form this mental conception, is to go through all the practical possibilities which Pallor Mala would afford him by its various movements and turns. In our world these various relations are found by the most simple observations; but a plane-being could only acquire them by considerable labour. And if he determined to obtain a knowledge of the physical existence of a higher world than his own, he must pass through such discipline.

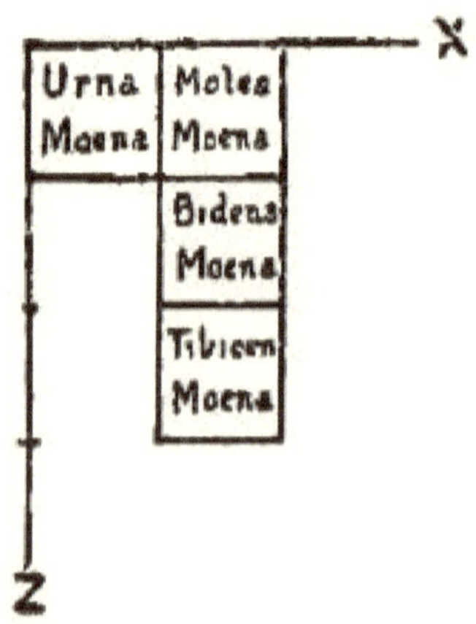


Fig. 1

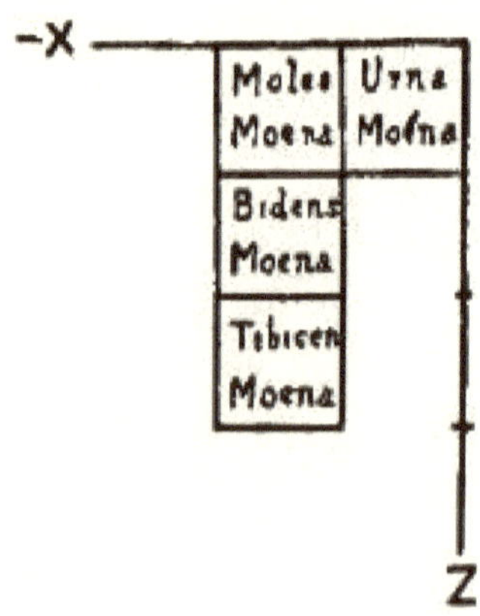


Fig. 16.

We will see what change could be introduced into the shapes he builds by the movements, which he does not know in his world. Let us build up this shape with the cubes of the Block: Urna Mala, Moles Mala, Bidens Mala, Tibicen Mala. To the plane-being this shape would be the slabs, Urna Moena, Moles Moena, Bidens Moena, Tibicen Moena (Fig. 15). Now let the Block be turned round the Z axis, so that it goes past the position, in which the Alvus sides enter the vertical plane. Let it move until, passing through the plane, the same Moena sides come in again. The mass of the Block will now have cut through the plane and be on the near side of it towards us; but the Moena faces only will be on the plane-being's side of it. The diagram (Fig. 16) shows what he will see, and it will seem to him similar to the first shape (Fig. 15) in every respect except its disposition with regard to the Z axis. It lies in the direction -X, opposite to that of the first figure. However much he turn these two figures about in the plane, he cannot make one occupy the place of the other, part for part. Hence it appears that, if we turn the plane-being's figure about a line, it undergoes an operation which is to him quite mysterious. He cannot by any turn in his plane produce the change in the figure produced by us. A little observation will show that a plane-being can only turn round a point. Turning round a line is a process unknown to him. Therefore one of the elements in a plane-being's knowledge of a space higher than his own, will be the conception of a kind of turning which will change his solid bodies into their own images.

# CHAPTER VI.
# THE MEANS BY WHICH A PLANE-BEING WOULD ACQUIRE A CONCEPTION OF OUR FIGURES.

Take the Block of twenty-seven Mala cubes, and build up the following shape (Fig. 18):—

Urna Mala, Moles Mala, Plebs Mala, Pallor Mala, Mora Mala.

If this shape, passed through the vertical plane, the plane-being would perceive:—

(1) The squares Urna Moena and Moles Moena.

(2) The three squares Plebs Moena, Pallor Moena, Mora Moena,
and then the whole figure would have passed through his plane.

If the whole Block were turned round the Z axis till the Alvus sides entered, and the figure built up as it would be in that disposition of the cubes, the plane-being would perceive during its passage through the plane:—

(1) Urna Alvus;

(2) Moles Alvus, Plebs Alvus, Pallor Alvus, Mora Alvus, which would all enter on the left side of the Z axis.

Again, if the Block were turned round the X axis, the Syce side would enter, and the cubes appear in the following order:—

(1) Urna Syce, Moles Syce, Plebs Syce;

(2) Pallor Syce;

(3) Mora Syce.

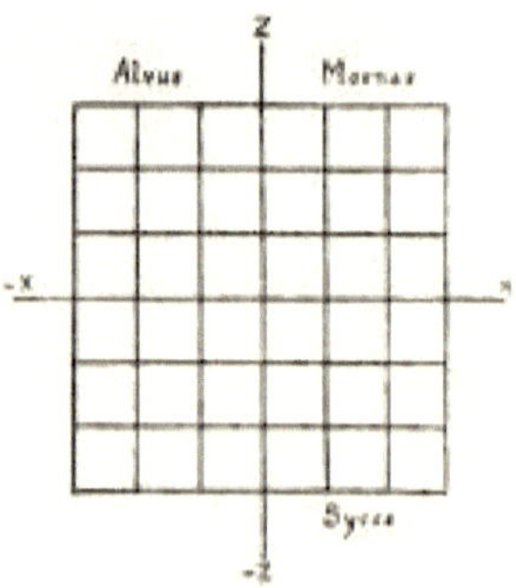

Fig. 17.

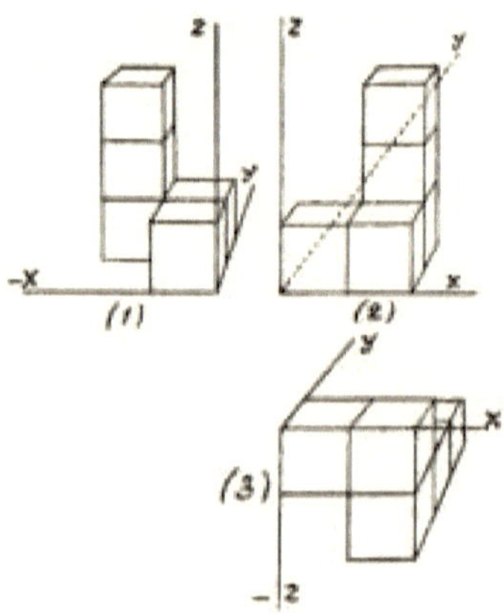

Fig. 18.

A comparison of these three sets of appearances would give the plane-being a full account of the figure. It is that which can produce these various appearances.

Let us now suppose a glass plate placed in front of the Block in its first position. On this plate let the axes X and Z be drawn. They divide the surface into four parts, to which we give the following names (Fig. 17):—

Z X = that quarter defined by the positive Z and positive X axis.

Z X = that quarter defined by the positive Z and negative X axis (which is called "Z negative X").

Z X = that quarter defined by the negative Z and negative X axis.

Z X = that quarter defined by the negative Z and positive X axis.

The Block appears in these different quarters or quadrants, as it is turned round the different axes. In Z X the Moenas appear, in Z X the Alvus faces, in Z X the Syces. In each quadrant are drawn nine squares, to receive the faces of the cubes when they enter. For instance, in Z X we have the Moenas of:—

Z

| Comes | Tibicen | Vestis |
|---|---|---|
| Ostrum | Bidens | Scena |
| Urna | Moles | Saltus |

X

And in Z X we have the Alvus of:—

Z

| Mars | Spicula | Comes |
|---|---|---|
| Ala | Uncus | Ostrum |
| Sector | Frenum | Urna |

-X

And in the Z X we have the Syces of:—

X

| Urna | Moles | Saltus |
|---|---|---|
| Frenum | Plebs | Sypho |
| Sector | Hama | Remus |

-Z

Now, if the shape taken at the beginning of this chapter be looked at through the glass, and the distance of the second and third walls of the shape behind the glass be considered of no account—that is, if they be treated as close up to the glass—we get a plane outline, which occupies the squares Urna Moena, Moles Moena, Bidens Moena, Tibicen Moena. This outline is called a projection of the figure. To see it like a plane-being, we should have to look down on it along the Z axis.

It is obvious that one projection does not give the shape. For instance, the square Bidens Moena might be filled by either Pallor or Cortis. All that a square in the room of Bidens Moena denotes, is that there is a cube somewhere in the Y, or unknown, direction from Bidens Moena. This view, just taken, we should call the front view in our space; we are then looking at it along the negative Y axis.

When the same shape is turned round on the Z axis, so as to be projected on the Z X quadrant, we have the squares—Urna Alvus, Frenum Alvus, Uncus Alvus, Spicula Alvus. When it is turned round the X axis, and projected on Z X, we have the squares, Urna Syce, Moles Syce, Plebs Syce, and no more. This is what is ordinarily called the ground plan; but we have set it in a different position from that in which it is usually drawn.

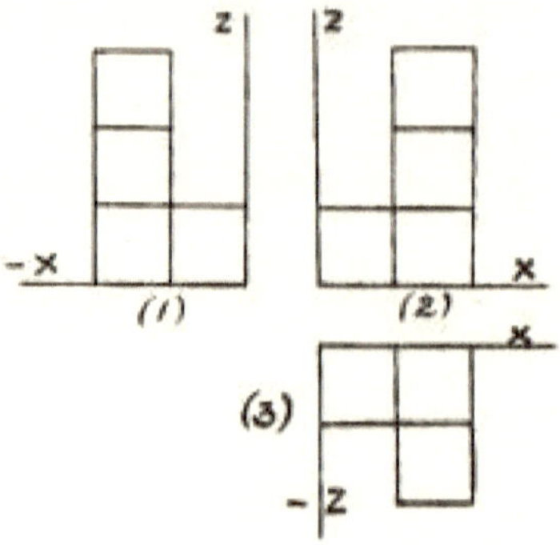


Fig. 19.

Now, the best method for a plane-being of familiarizing himself with shapes in our space, would be to practise the realization of them from their different projections in his plane. Thus, given the three projections just mentioned, he should be able to construct the figure from which they are derived. The projections (Fig. 19) are drawn below the perspective pictures of the shape (Fig. 18). From the front, or Moena view, he would conclude that the shape was Urna Mala, Moles Mala, Bidens Mala, Tibicen Mala; or instead of these, or also in addition to them, any of the cubes running in the Y direction from the plane. That is, from the Moena projection he might infer the presence of all the following cubes (the word Mala is omitted for brevity): Urna, Frenum, Sector, Moles, Plebs, Hama, Bidens, Pallor, Cortis, Tibicen, Mora, Merces.

Next, the Alvus view or projection might be given by the cubes (the word Mala being again omitted): Urna, Moles, Saltus, Frenum, Plebs, Sypho, Uncus, Pallor, Tergum, Spicula, Mora, Oliva. Lastly, looking at the ground plan or Syce view, he would infer the possible presence of Urna, Ostrum, Comes, Moles, Bidens, Tibicen, Plebs, Pallor, Mora.

Now, the shape in higher space, which is usually there, is that which is common to all these three appearances. It can be determined, therefore, by rejecting those cubes which are not present in all three lists of cubes possible from the projections. And by this process the plane-being could arrive at the enumeration of the cubes which belong to the shape of which he had the projections. After a time, when he had experience of the cubes (which, though invisible to him as wholes, he could see part by part in turn entering his space), the projections would have more meaning to him, and he might comprehend the shape they expressed fragmentarily in his space. To practise the realization from projections, we should proceed in this way. First, we should think of the possibilities involved in the Moena view, and build them up in cubes before us. Secondly, we should build up the cubes possible from the Alvus view. Again, taking the shape at the beginning of the chapter, we should find that the shape of the Alvus possibilities intersected that of the Moena possibilities in Urna, Moles, Frenum, Plebs, Pallor, Mora; or, in other words, these cubes are common to both. Thirdly, we should build up the Syce possibilities, and, comparing their shape with those of the Moena and Alvus projections, we should find Urna, Moles, Plebs, Pallor, Mora, of the Syce view coinciding with the same cubes of the other views, the only cube present in the intersection of the Moena and Alvus possibilities, and not present in the Syce view, being Frenum.

The determination of the figure denoted by the three projections, may be more easily effected by treating each projection as an indication of what cubes are to be cut away. Taking the same shape as before, we have in the Moena projection Urna, Moles, Bidens, Tibicen; and the possibilities from them are Urna, Frenum, Sector, Moles, Plebs, Hama, Bidens, Pallor, Cortis, Tibicen, Mora, Merces. This may aptly be called the Moena solution. Now, from the Syce projection, we learn at once that those cubes, which in it would produce Frenum, Sector, Hama, Remus, Sypho, Saltus, are not in the shape. This absence of Frenum and Sector in the Syce view proves that their presence in the Moena solution is superfluous. The absence of Hama removes the possibility of Hama, Cortis, Merces. The absence of Remus, Sypho, Saltus, makes no difference, as neither they nor any of their Syce possibilities are present in the Moena solution. Hence, the result of comparison of the Moena and Syce projections and possibilities is the shape: Urna, Moles, Plebs, Bidens, Pallor, Tibicen, Mora. This may be aptly called the Moena-Syce solution. Now, in the Alvus projection we see that Ostrum, Comes, Sector, Ala, and Mars are absent. The absence of Sector, Ala, and Mars has no effect on our Moena-Syce solution; as it does not contain any of their Alvus possibilities. But the absence of Ostrum and Comes proves that in the Moena-Syce solution Bidens and Tibicen are superfluous, since their presence in the original shape would give Ostrum and Comes in the Alvus projection. Thus we arrive at the Moena-Alvus-Syce solution, which gives us the shape: Urna, Moles, Plebs, Pallor, Mora.

It will be obvious on trial that a shape can be instantly recognised from its three projections, if the Block be thoroughly well known in all three positions. Any difficulty in the realization of the shapes comes from the arbitrary habit of associating the cubes with some one direction in which they happen to go with regard to us. If we remember Ostrum as above Urna, we are not remembering the Block, but only one particular relation of the Block to us. That position of Ostrum is a fact as much related to ourselves as to the Block. There is, of course, some information about the Block implied in that position; but it is so mixed with information about ourselves as to be ineffectual knowledge of the Block. It is of the highest importance to enter minutely into all the details of solution written above. For, corresponding to every operation necessary to a plane-being for the comprehension of our world, there is an operation, with which we have to become familiar, if in our turn we would enter into some comprehension of a world higher than our own. Every cube of the Block ought to be thoroughly known in all its relations. And the Block must be regarded, not as a formless mass out of which shapes can be made, but as the sum of all possible shapes, from which any one we may choose is a selection. In fact, to be familiar with the Block, we ought to know every shape that could be made by any selection of its cubes; or, in other words, we ought to make an exhaustive study of it. In the Appendix is given a set of exercises in the use of these names (which form a language of shape), and in various kinds of projections. The projections studied in this chapter are not the only, nor the most natural, projections by which a plane-being would study higher space. But they suffice as an illustration of our present purpose. If the reader will go through the exercises in the Appendix, and form others for himself, he will find every bit of manipulation done will be of service to him in the comprehension of higher space.

There is one point of view in the study of the Block, by means of slabs, which is of some interest. The cubes of the Block, and therefore also the representative slabs of their faces, can be regarded as forming rows and columns. There are three sets of them. If we take the Moena view, they represent the views of the three walls of the Block, as they pass through the plane. To form the Alvus view, we only have to rearrange the slabs, and form new sets. The first new set is formed by taking the first, or left-hand, column of each of the Moena sets. The second Alvus set is formed by taking the second or middle columns of the three Moena sets. The third will consist of the remaining or right-hand columns of the Moenas.

Similarly, the three Syce sets may be formed from the three horizontal rows or floors of the Moena sets.

Hence, it appears that the plane-being would study our space by taking all the possible combinations of the corresponding rows and columns. He would break up the first three sets into other sets, and the study of the Block would practically become to him the study of these various arrangements.

# CHAPTER VII.
# FOUR-SPACE: ITS REPRESENTATION IN THREE-SPACE.

We now come to the essential difficulty of our task. All that has gone before is preliminary. We have now to frame the method by which we shall introduce through our space-figures the figures of a higher space. When a plane-being studies our shapes of cubes, he has to use squares. He is limited at the outset. A cube appears to him as a square. On Model 1 we see the various squares as which the cube can appear to him. We suppose the plane-being to look from the extremity of the Z axis down a vertical plane. First, there is the Moena square. Then there is the square given by a section parallel to Moena, which he recognises by the variation of the bounding lines as soon as the cube begins to pass through his plane. Then comes the Murex square. Next, if the cube be turned round the Z axis and passed through, he sees the Alvus and Proes squares and the intermediate section. So too with the Syce and Mel squares and the section between them.

Now, dealing with figures in higher space, we are in an analogous position. We cannot grasp the element of which they are composed. We can conceive a cube; but that which corresponds to a cube in higher space is beyond our grasp. But the plane-being was obliged to use two-dimensional figures, squares, in arriving at a notion of a three-dimensional figure; so also must we use three-dimensional figures to arrive at the notion of a four-dimensional. Let us call the figure which corresponds to a square in a plane and a cube in our space, a tessaract. Model 1 is a cube. Let us assume a tessaract generated from it. Let us call the tessaract Urna. The generating cube may then be aptly called Urna Mala. We may use cubes to represent parts of four-space, but we must always remember that they are to us, in our study, only what squares are to a plane-being with respect to a cube.

Let us again examine the mode in which a plane-being represents a Block of cubes with slabs. Take Block 1 of the 81 Set of cubes. The plane-being represents this by nine slabs, which represent the Moena face of the block. Then, omitting the solidity of these first nine cubes, he takes another set of nine slabs to represent the next wall of cubes. Lastly, he represents the third wall by a third set, omitting the solidity of both second and third walls. In this manner, he evidently represents the extension of the Block upwards and sideways, in the Z and X directions; but in the Y direction he is powerless, and is compelled to represent extension in that direction by setting the three sets of slabs alongside in his plane. The second and third sets denote the height and breadth of the respective walls, but not their depth or thickness. Now, note that the Block extends three inches in each of the three directions. The plane-being can represent two of these dimensions on his plane; but the unknown direction he has to represent by a repetition of his plane figures. The cube extends three inches in the Y direction. He has to use 3 sets of slabs.

The Block is built up arbitrarily in this manner: Starting from Urna Mala and going up, we come to a Brown cube, and then to a Light-blue cube. Starting from Urna Mala and going right, we come to an Orange and a Fawn cube. Starting from Urna Mala and going away from us, we come to a Blue and a Buff cube. Now, the plane-being represents the Brown and Orange cubes by squares lying next to the square which represents Urna Mala. The Blue cube is as close as the Brown cube to Urna Mala, but he can find no place in the plane where he can place a Blue square so as to show this co-equal proximity of both cubes to the first. So he is forced to put a Blue square anywhere in his plane and say of it: "This Blue square represents what I should arrive at, if I started from Urna Mala and moved away, that is in the Y or unknown direction." Now, just as there are three cubes going up, so there are three going away. Hence, besides the Blue square placed anywhere on the plane, he must also place a Buff square beyond it, to show that the Block extends as far away as it does upwards and sideways. (Each cube being a different colour, there will be as many different colours of squares as of cubes.) It will easily be seen that not only the Gold square, but also the Orange and every other square in the first set of slabs must have two other squares set somewhere beyond it on the plane to represent the extension of the Block away, or in the unknown Y direction.

Coming now to the representation of a four-dimensional block, we see that we can show only three dimensions by cubic blocks, and that the fourth can only be represented by repetitions of such blocks. There must be a certain amount of arbitrary naming and colouring. The colours have been chosen as now stated. Take the first Block of the 81 Set. We are familiar with its colours, and they can be found at any time by reference to Model 1. Now, suppose the Gold cube to represent what we can see in our space of a Gold tessaract; the other cubes of Block 1 give the colours of the tessaracts which lie in the three directions X, Y, and Z from the Gold one. But what is the colour of the tessaract which lies next to the Gold in the unknown direction, W? Let us suppose it to be Stone in colour. Taking out Block 2 of the 81 Set and arranging it on the pattern of Model 9, we find in it a Stone cube. But, just as there are three tessaracts in the X, Y, and Z directions, as shown by the cubes in Block 1, so also must there be three tessaracts in the unknown direction, W. Take Block 3 of the 81 Set. This Block can be arranged on the pattern of Model 2. In it there is a Silver cube where the Gold cube lies in Block 1. Hence, we may say, the tessaract which comes next to the Stone one in the unknown direction from the Gold, is of a Silver colour. Now, a cube in all these cases represents a tessaract. Between the Gold and Stone cubes there is an inch in the unknown direction. The Gold tessaract is supposed to be Gold throughout in all four directions, and so also is the Stone. We may imagine it in this way. Suppose the set of three tessaracts, the Gold, the Stone, and the Silver to move through our space at the rate of an inch a minute. We should first see the Gold cube which would last a minute, then the Stone cube for a minute, and lastly the Silver cube a minute. (This is precisely analogous to the appearance of passing cubes to the plane-being as successive squares lasting a minute.) After that, nothing would be visible.

Now, just as we must suppose that there are three tessaracts proceeding from the Gold cube in the unknown direction, so there must be three tessaracts extending in the unknown direction from every one of the 27 cubes of the first Block. The Block of 27 cubes is not a Block of 27 tessaracts, but it represents as much of them as we can see at once in our space; and they form the first portion or layer (like the first wall of cubes to the plane-being) of a set of eighty-one tessaracts, extending to equal distances in all four directions. Thus, to represent the whole Block of tessaracts there are 81 cubes, or three Blocks of 27 each.

Now, it is obvious that, just as a cube has various plane boundaries, so a tessaract has various cube boundaries. The cubes of the tessaract, which we have been regarding, have been those containing the X, Y, and Z directions, just as the plane-being regarded the Moena faces containing the X and Z directions. And, as long as the tessaract is unchanged in its position with regard to our space, we can never see any portion of it which is in the unknown direction. Similarly, we saw that a plane-being could not see the parts of a cube which went in the third direction, until the cube was turned round one of its edges. In order to make it quite clear what parts of a cube came into the plane, we gave distinct names to them. Thus, the squares containing the Z and X directions were called Moena and Murex; those containing the Z and Y, Alvus and Proes; and those the X and Y, Syce and Mel. Now, similarly with our four axes, any three will determine a cube. Let the tessaract in its normal position have the cube Mala determined by the axes Z, X, Y. Let the cube Lar be that which is determined by X, Y, W, that is, the cube which, starting from the X Y plane, stretches one inch in the unknown or W direction. Let Vesper be the cube determined by Z, Y, W, and Pluvium by Z, X, W. And let these cubes have opposite cubes of the following names:

| | | |
|---|---|---|
| Mala | has | Margo |
| Lar | „ | Velum |
| Vesper | „ | Idus |

Pluvium „ Tela

Another way of looking at the matter is this. When a cube is generated from a square, each of the lines bounding the square becomes a square, and the square itself becomes a cube, giving two squares in its initial and final positions. When a cube moves in the new and unknown direction, each of its planes traces a cube and it generates a tessaract, giving in its initial and final positions two cubes. Thus there are eight cubes bounding the tessaract, six of them from the six plane sides and two from the cube itself. These latter two are Mala and Margo. The cubes from the six sides are: Lar from Syce, Velum from Mel, Vesper from Alvus, Idus from Proes, Pluvium from Moena, Tela from Murex. And just as a plane-being can only see the squares of a cube, so we can only see the cubes of a tessaract. It may be said that the cube can be pushed partly through the plane, so that the plane-being sees a section between Moena and Murex. Similarly, the tessaract can be pushed through our space so that we can see a section between Mala and Margo.

There is a method of approaching the matter, which settles all difficulties, and provides us with a nomenclature for every part of the tessaract. We have seen how by writing down the names of the cubes of a block, and then supposing that their number increases, certain sets of the names come to denote points, lines, planes, and solid. Similarly, if we write down a set of names of tessaracts in a block, it will be found that, when their number is increased, certain sets of the names come to denote the various parts of a tessaract.

For this purpose, let us take the 81 Set, and use the cubes to represent tessaracts. The whole of the 81 cubes make one single tessaractic set extending three inches in each of the four directions. The names must be remembered to denote tessaracts. Thus, Corvus is a tessaract which has the tessaracts Cuspis and Nugæ to the right, Arctos and Ilex above it, Dos and Cista away from it, and Ops and Spira in the fourth or unknown direction from it. It will be evident at once, that to write these names in any representative order we must adopt an arbitrary system. We require them running in four directions; we have only two on paper. The X direction (from left to right) and the Y (from the bottom towards the top of the page) will be assumed to be truly represented. The Z direction will be symbolized by writing the names in floors, the upper floors always preceding the lower. Lastly, the fourth, or W, direction (which has to be symbolized in three-dimensional space by setting the solids in an arbitrary position) will be signified by writing the names in blocks, the name which stands in any one place in any block being next in the W direction to that which occupies the same position in the block before or after it. Thus, Ops is written in the same place in the Second Block, Spira in the Third Block, as Corvus occupies in the First Block.

Since there are an equal number of tessaracts in each of the four directions, there will be three floors Z when there are three X and Y. Also, there will be three Blocks W. If there be four tessaracts in each direction, there will be four floors Z, and four blocks W. Thus, when the number in each direction is enlarged, the number of blocks W is equal to the number of tessaracts in each known direction.

On pp. 136, 137 were given the names as used for a cubic block of 27 or 64. Using the same and more names for a tessaractic Set, and remembering that each name now represents, not a cube, but a tessaract, we obtain the following nomenclature, the order in which the names are written being that stated above:

THIRD BLOCK.

Solia Livor Talus
Lensa Lares Calor
Felis Tholus Passer

Upper Floor.

Lixa Portica Vena
Crux Margo Sal
Pagus Silex Onager

Middle Floor.

Lower Floor

Panax Mensura Mugil

oor.

| | | |
|---|---|---|
| Opex | Lappa | Mappa |
| Spira | Luca | Ancilla |

SECOND BLOCK.

| | | |
|---|---|---|
| Orsa | Mango | Libera |
| Creta | Velum | Meatus |
| Lucta | Limbus | Pator |

Upper - Floor.

| | | |
|---|---|---|
| Camoena | Tela | Orca |
| Vesper | Tessaract | Idus |
| Pagina | Pluvium | Pactum |

Middle - Floor.

| | | |
|---|---|---|
| Lis | Lorica | Offex |
| Lua | Lar | Olla |
| Ops | Lotus | Limus |

Lower - Floor.

FIRST BLOCK.

Upper - Floor.

| | | |
|---|---|---|
| Olus | Semita | Lama |
| Via Il | Mel C | Iter S |

e all o
x is rs

| | | | |
|---|---|---|---|
| Middle Floor. | Buccina | Murex | Daps |
| | Alvus | Mala | Proes |
| | Arctos | Moena | Far |

| | | | |
|---|---|---|---|
| Lower Floor. | Cista | Cadus | Crus |
| | Dos | Syce | Bolus |
| | Corvus | Cuspis | Nugæ |

It is evident that this set of tessaracts could be increased to the number of four in each direction, the names being used as before for the cubic blocks on pp. 136, 137, and in that case the Second Block would be duplicated to make the four blocks required in the unknown direction. Comparing such an 81 Set and 256 Set, we should find that Cuspis, which was a single tessaract in the 81 Set became two tessaracts in the 256 Set. And, if we introduced a larger number, it would simply become longer, and not increase in any other dimension. Thus, Cuspis would become the name of an edge. Similarly, Dos would become the name of an edge, and also Arctos. Ops, which is found in the Middle Block of the 81 Set, occurs both in the Second and Third Blocks of the 256 Set; that is, it also tends to elongate and not extend in any other direction, and may therefore be used as the name of an edge of a tessaract.

Looking at the cubes which represent the Syce tessaracts, we find that, though they increase in number, they increase only in two directions; therefore, Syce may be taken to signify a square. But, looking at what comes from Syce in the W direction, we find in the Middle Block of the 81 Set one

Lar, and in the Second and Third Blocks of the 256 Set four Lars each. Hence, Lar extends in three directions, X, Y, W, and becomes a cube. Similarly, Moena is a plane; but Pluvium, which proceeds from it, extends not only sideways and upwards like Moena, but in the unknown direction also. It occurs in both Middle Blocks of the 256 Set. Hence, it also is a cube. We have now considered such parts of the Sets as contain one, two, and three dimensions. But there is one part which contains four. It is that named Tessaract. In the 256 Set there are eight such cubes in the Second, and eight in the Third Block; that is, they extend Z, X, Y, and also W. They may, therefore, be considered to represent that part of a tessaract or tessaractic Set, which is analogous to the interior of a cube.

The arrangement of colours corresponding to these names is seen on Model 1 corresponding to Mala, Model 2 to Margo, and Model 9 to the intermediate block.

When we take the view of the tessaract with which we commenced, and in which Arctos goes Z, Cuspis X, Dos Y, and Ops W, we see Mala in our space. But when the tessaract is turned so that the Ops line goes -X, while Cuspis is turned W, the other two remaining as they were, then we do not see Mala, but that cube which, in the original position of the tessaract, contains the Z, Y, W, directions, that is, the Vesper cube.

A plane-being may begin to study a block of cubes by their Syce squares; but if the block be turned round Dos, he will have Alvus squares in his space, and he must then use them to represent the cubic Block. So, when the tessaractic Set is turned round, Mala cubes leave our space, and Vespers enter.

There are two ways which can be followed in studying the Set of tessaracts.

I. Each tessaract of one inch every way can be supposed to be of the same colour throughout, so that, whichever way it be turned, whichever of its edges coincide with our known axes, it appears to us as a cube of one uniform colour. Thus, if Urna be the tessaract, Urna Mala would be a Gold cube, Urna Vesper a Gold cube, and so on. This method is, for the most part, adopted in the following pages. In this case, a whole Set of $4 \times 4 \times 4 \times 4$ tessaracts would in colours resemble a set composed of four cubes like Models 1, 9, 9, and 2. But, when any question about a particular tessaract has to be settled, it is advantageous, for the sake of distinctness, to suppose it coloured in its different regions as the whole set is coloured.

II. The other plan is, to start with the cubic sides of the inch tessaract, each coloured according to the scheme of the Models 1 to 8. In this case, the lines, if shown at all, should be very thin. For, in fact, only the faces would be seen, as the width of the lines would only be equal to the thickness of our matter in the fourth dimension, which is indistinguishable to the senses. If such completely coloured cubes be used, less error is likely to creep in; but it is a disadvantage that each cube in the several blocks is exactly like the others in that block. If the reader make such a set to work with for a time, he will gain greatly, for the real way of acquiring a sense of higher space is to obtain those experiences of the senses exactly, which the observation of a four-dimensional body would give. These Models 1-8 are called sides of the tessaract.

To make the matter perfectly clear, it is best to suppose that any tessaract or set of tessaracts which we examine, has a duplicate exactly similar in shape and arrangement of parts, but different in their colouring. In the first, let each tessaract have one colour throughout, so that all its cubes, apprehended in turn in our space, will be of one and the same colour. In the duplicate, let each tessaract be so coloured as to show its different cubic sides by their different colours. Then, when we have it turned to us in different aspects, we shall see different cubes, and when we try to trace the contacts of the tessaracts with each other, we shall be helped by realizing each part of every tessaract in its own colour.

# CHAPTER VIII.
# REPRESENTATION OF FOUR-SPACE BY NAME. STUDY OF TESSARACTS.

We have now surveyed all the preliminary ground, and can study the masses of tessaracts without obscurity.

We require a scaffold or framework for this purpose, which in three dimensions will consist of eight cubic spaces or octants assembled round one point, as in two dimensions it consisted of four squares or quadrants round a point.

These eight octants lie between the three axes Z, X, Y, which intersect at the given point, and can be named according to their positions between the positive and negative directions of those axes. Thus the octant Z, X, Y, is that which is contained by the positive portions of all three axes; the octant Z, X, Y, that which is to the left of Z, X, Y, and between the positive parts of Z and Y and the negative of X. To illustrate this quite clearly, let us take the eight cubes—Urna, Moles, Plebs, Frenum, Uncus, Pallor, Bidens, Ostrum—and place them in the eight octants. Let them be placed round the point of intersection of the axes; Pallor Corvus, Plebs Ilex, etc., will be at that point. Their positions will then be:—

| | | | |
|---|---|---|---|
| Urna | in the | Octant | Z X Y |
| Moles | ,, | ,, | Z X Y |
| Plebs | ,, | ,, | Z X Y |
| Frenum | ,, | ,, | Z X Y |
| Uncus | ,, | ,, | Z X Y |
| Pallor | ,, | ,, | Z X Y |
| Bidens | ,, | ,, | Z X Y |
| Ostrum | ,, | ,, | Z X Y |

The names used for the cubes, as they are before us, are as follows:—

THIRD BLOCK.

| | | | |
|---|---|---|---|
| Third | Arcus Mala | Ovis Mala | Portio Mala |
| Floor. | Laurus Mal | Tigris Mal | Segmen Mal |

a a a

Axis Mala Troja Mala Aries Mala

Postis Mala Clipeus Mala Tabula Mala

Orcus Mala Lacerta Mala Testudo Mala

Verbum Mala Luctus Mala Anguis Mala

Second-Floor.

Telum Mala Nepos Mala Angusta Mala

Polus Mala Penates Mala Vulcan Mala

First-Floor.

Cervix Mala Securis Mala Vinculum Mala

SECOND BLOCK.

Th - A V Pl

- ird Floor.
  - ra Mala
  - omer Mala
  - uma Mala
  - Praeda Mala
  - Sacerdos Mala
  - Hydra Mala
  - Cortex Mala
  - Mica Mala
  - Flagellum Mala
- Second Floor.
  - Pilum Mala
  - Glans Mala
  - Colus Mala
  - Ocrea Mala
  - Tessera Mala
  - Domitor Mala
  - Cardo Mala
  - Cudo Mala
  - Malleus Mala
- First Floor.
  - Agmen Mala
  - Lacus Mala
  - Arvus Mala
  - Crates M
  - Cura
  - Limen M

ala Mala ala

Thyrsus Mala Vitta Mala Sceptrum Mala

FIRST BLOCK.

Mars Mala Merces Mala Tyro Mala

Spicula Mala Mora Mala Oliva Mala

Comes Mala Tibicen Mala Vestis Mala

Third Floor.

Ala Mala Cortis Mala Aer Mala

Second Floor.

Uncus Mala Pallor Mala Tergum Mala

Ostrum Mala Bidens Mala Scena Mala

| | | | |
|---|---|---|---|
| First Floor. | Sector Mala | Hama Mala | Remus Mala |
| | Frenum Mala | Plebs Mala | Sypho Mala |
| | Urna Mala | Moles Mala | Saltus Mala |

Their colours can be found by reference to the Models 1, 9, 2, which correspond respectively to the First, Second, and Third Blocks. Thus, Urna Mala is Gold; Moles, Orange; Saltus, Fawn; Thyrsus, Stone; Cervix, Silver. The cubes whose colours are not shown in the Models, are Pallor Mala, Tessera Mala, and Lacerta Mala, which are equivalent to the interiors of the Model cubes, and are respectively Light-buff, Wooden, and Sage-green. These 81 cubes are the cubic sides and sections of the tessaracts of an 81 tessaractic Set, which measures three inches in every direction. We suppose it to pass through our space. Let us call the positive unknown direction Ana (*i.e.*, +W) and the negative unknown direction Kata (-W). Then, as the whole tessaract moves Kata at the rate of an inch a minute, we see first the First Block of 27 cubes for one minute, then the Second, and lastly the Third, each lasting one minute.

Now, when the First Block stands in the normal position, the edges of the tessaract that run from the Corvus corner of Urna Mala, are: Arctos in Z, Cuspis in X, Dos in Y, Ops in W. Hence, we denote this position by the following symbol:—

$$\begin{matrix} Z & X & Y & W \\ a & c & d & o \end{matrix}$$

where *a* stands for Arctos, *c* for Cuspis, *d* for Dos, and *o* for Ops, and the other letters for the four axes in space. *a*, *c*, *d*, *o* are the axes of the tessaract, and can take up different directions in space with regard to us.

Let us now take a smaller four-dimensional set. Of the 81 Set let us take the following:—

Z X Y W

*a c d o*

SECOND BLOCK.

Ocrea Mala — Tessera Mala
Cardo Mala — Cudo Mala

Crates Mala — Cura Mala
Thyrsus Mala — Vitta Mala

Second Floor.

First Floor.

FIRST BLOCK.

Uncus Mala — Pallor Mala
Ostrum Mala — Bidens Mala

Second Floor.

Frenum Mala — Plebs Mala
Urna Mala — Moles Mala

First Floor.

Let the First Block be put up before us in Z X Y, (Urna Corvus is at the junction of our axes Z X Y). The Second Block is now one inch distant in the unknown direction; and, if we suppose the tessaractic Set to move through our space at the rate of one inch a minute, the Second will enter in one minute, and replace the first. But, instead of this, let us suppose the tessaracts to turn so that Ops, which now goes W, shall go -X. Then we can see in our space that cubic side of each tessaract which is contained by the lines Arctos, Dos, and Ops, the cube Vesper; and we shall no longer have the Mala sides but the Vesper sides of the tessaractic Set in our space. We will now build it up in its Vesper view (as we built up the cubic Block in its Alvus view). Take the Gold cube, which now means Urna Vesper, and place it on the left hand of its former position as Urna Mala, that is, in the octant Z X Y. Thyrsus Vesper, which previously lay just beyond Urna Vesper in the unknown direction, will now lie just beyond it in the -X direction, that is, to the left of it. The tessaractic Set is now in the position

Z

*a*

X

$\bar{o}$

Y
*d*

W
*c*

(the minus sign over the *o* meaning that Ops runs in the negative direction), and its Vespers lie in the following order:—

SECOND BLOCK.

Second Floor.
Tessara Pallor
Cudo Bidens

First Floor.
Cura Plebs
Vitta Moles

FIRST BLOCK.

Second Floor.
Ocrea Uncus
Cardo Ostrum

First Floor.
Crates Frenum
Thyrsus Urna

The name Vesper is left out in the above list for the sake of brevity, but should be used in studying the positions.

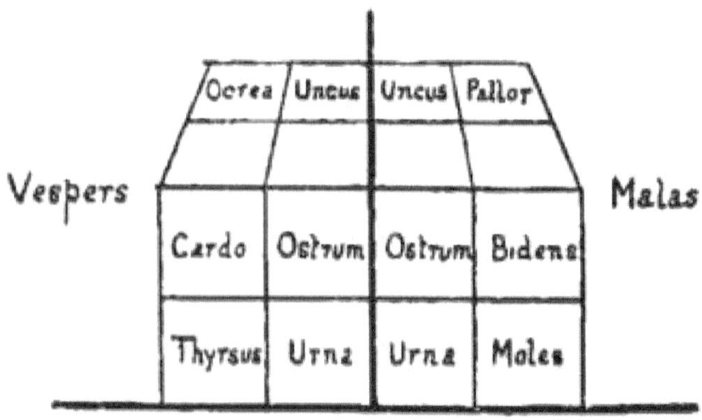


Fig. 20.

On comparing the two lists of the Mala view and Vesper view, it will be seen that the cubes presented in the Vesper view are new sides of the tessaract, and that the arrangement of them is different from that in the Mala view. (This is analogous to the changes in the slabs from the Moena to Alvus view of the cubic Block.) Of course, the Vespers of all these tessaracts are not visible at once in our space, any more than are the Moenas of all three walls of a cubic Block to a plane-being. But if the tessaractic Set be supposed to move through space in the unknown direction at the rate of an inch a minute, the Second Block will present its Vespers after the First Block has lasted a minute. The relative position of the Mala Block and the Vesper Block may be represented in our space as in the diagram, Fig. 20. But it must be distinctly remembered that this arrangement is quite conventional, no more real than a plane-being's symbolization of the Moena Wall and the Alvus Wall of the cubic Block by the arrangement of their Moena and Alvus faces, with the solidity omitted, along one of his known directions.

The Vespers of the First and Second Blocks cannot be in our space simultaneously, any more than the Moenas of all three walls in plane space. To render their simultaneous presence possible, the cubic or tessaractic Block or Set must be broken up, and its parts no longer retain their relations. This fact is of supreme importance in considering higher space. Endless fallacies creep in as soon as it is forgotten that the cubes are merely representative as the slabs were, and the positions in our space merely conventional and symbolical, like those of the slabs along the plane. And these fallacies are so much fostered by again symbolizing the cubic symbols and their symbolical positions in perspective drawings or diagrams, that the reader should surrender all hope of learning space from this book or the drawings alone, and work every thought out with the cubes themselves.

If we want to see what each individual cube of the tessaractic faces presented to us in the last example is like, we have only to consider each of the Malas similar in its parts to Model 1, and each of the Vespers to Model 5. And it must always be remembered that the cubes, though used to represent both Mala and Vesper faces of the tessaract, mean as great a difference as the slabs used for the Moena and Alvus faces of the cube.

If the tessaractic Set move Kata through our space, when the Vesper faces are presented to us, we see the following parts of the tessaract Urna (and, therefore, also the same parts of the other tessaracts):

(1) Urna Vesper, which is Model 5.

(2) A parallel section between Urna Vesper and Urna Idus, which is Model 11.

(3) Urna Idus, which is Model 6.

When Urna Idus has passed Kata our space, Moles Vesper enters it; then a section between Moles Vesper and Moles Idus, and then Moles Idus. Here we have evidently observed the tessaract more minutely; as it passes Kata through our space, starting on its Vesper side, we have seen the parts which would be generated by Vesper moving along Cuspis—that is Ana.

Two other arrangements of the tessaracts have to be learnt besides those from the Mala and Vesper aspect. One of them is the Pluvium aspect. Build up the Set in Z X Y, letting Arctos run Z, Cuspis X, and Ops Y. In the common plane Moena, Urna Pluvium coincides with Urna Mala, though they cannot be in our space together; so too Moles Pluvium with Moles Mala, Ostrum Pluvium with Ostrum Mala. And lying towards us, or Y, is now that tessaract which before lay in the W direction from Urna, viz., Thyrsus. The order will therefore be the following (a star denotes the cube whose corner is at point of intersection of the axes, and the name Pluvium must be understood to follow each of the names):

Z X Y W

*a c ō d*

SECOND BLOCK.

Uncus Pallor
Ocrea Tessera

Frenum Plebs
Crates Cura

Second Floor.

First Floor.

Second Floor.

FIRST BLOCK.

Ostrum Bidens
Cardo Cudo

First Floor.

* Urna Moles
Thyrsus Vitta

Thus the wall of cubes in contact with that wall of the Mala position which contains the Urna, Moles, Ostrum, and Bidens Malas, is a wall composed of the Pluviums of Urna, Moles, Ostrum, and Bidens. The wall next to this, and nearer to us, is of Thyrsus, Vitta, Cardo, Cudo, Pluviums. The Second Block is one inch out of our Space, and only enters it if the Block moves Kata. Model 7 shows the Pluvium cube; and each of the cubes of the tessaracts seen in the Pluvium position is a Pluvium. If the tessaractic Set moved Kata, we would see the Section between Pluvium and Tela for all but a minute; and then Tela would enter our space, and the Tela of each tessaract would be seen. Model 12 shows the Section from Pluvium to Tela. Model 8 is Tela. Tela only lasts for a flash, as it has only the minutest magnitude in the unknown or Ana direction. Then, Frenum Pluvium takes the place of Urna Tela; and, when it passes through, we see a similar section between Frenum Pluvium and Frenum Tela, and lastly Frenum Tela. Then the tessaractic Set passes out, or Kata, our space. A similar process takes place with every other tessaract, when the Set of tessaracts moves through our space.

There is still one more arrangement to be learnt. If the line of the tessaract, which in the Mala position goes Ana, or W, be changed into the Z or downwards direction, the tessaract will then appear in our space below the Mala position; and the side presented to us will not be Mala, but that which contains the lines Dos, Cuspis, and Ops. This side is Model 3, and is called Lar. Underneath the place which was occupied by Urna Mala, will come Urna Lar; under the place of Moles Mala, Moles Lar; under the place of Frenum Mala, Frenum Lar. The tessaract, which in the Mala position was an inch out of our space Ana, or W, from Urna Mala, will now come into it an inch downwards, or Z, below Urna Mala, with its Lar presented to us; that is, Thyrsus Lar will be below Urna Lar. In the whole arrangement of them written below, the highest floors are written first, for now they stretch downwards instead of upwards. The name Lar is understood after each.

Z X Y W

*ō c d a*

SECOND BLOCK.

Uncus Pallor
Ostrum Bidens

Ocrea Tessera
Cardo Cudo

Second Floor. First Floor.

FIRST BLOCK.

Frenum Plebs
* Urna Moles

Second Floor.

First Floor.

Crates Cura
Thyrsus Vitta

Here it is evident that what was the lower floor of Malas, Urna, Moles, Plebs, Frenum, now appears as if carried downwards instead of upwards, Lars being presented in our space instead of Malas. This Block of Lars is what we see of the tessaract Set when the Arctos line, which in the Mala position goes up, is turned into the Ana, or W, direction, and the Ops line comes in downwards.

The rest of the tessaracts, which consists of the cubes opposite to the four treated above, and of the tessaractic space between them, is all Ana in our space. If the tessaract be moved through our space—for instance, when the Lars are present in it—we see, taking Urna alone, first the section between Urna Lar and Urna Velum (Model 10), and then Urna Velum (Model 4), and similarly the sections and Velums of each tessaract in the Set. When the First Block has passed Kata our space, Ostrum Lar enters; and the Lars of the Second Block of tessaracts occupy the places just vacated by the Velums of the First Block. Then, as the tessaractic Set moves on Kata, the sections between Velums and Lars of the Second Block of tessaracts enter our space, and finally their Velums. Then the whole tessaractic Set disappears from our space.

When we have learnt all these aspects and passages, we have experienced some of the principal features of this small Set of tessaracts.

# CHAPTER IX.
# FURTHER STUDY OF TESSARACTS.

When the arrangement of a small set has been mastered, the different views of the whole 81 Set should be learnt. It is now clear to us that, in the list of the names of the eighty-one tessaracts given above, those which lie in the W direction appear in different blocks, while those that lie in the Z, X, or Y directions can be found in the same block. Therefore, from the arrangement given, which is denoted by

Z
*a*

X
*c*

Y
*d*

W
*o*

or more briefly by *a c d o*, we can form any other arrangement.

To confirm the meaning of the symbol *a c d o* for position, let us remember that the order of the axes known in our space will invariably be Z X Y, and the unknown direction will be stated last, thus: Z X Y W. Hence, if we write *a ō d c*, we know that the position or aspect intended is that in which Arctos (*a*) goes Z, Ops (*ō*) negative X, Dos (*d*) Y, and Cuspis (*c*) W. And such an arrangement can be made by shifting the nine cubes on the left side of the First Block of the eighty-one tessaracts, and putting them into the Z X Y octant, so that they just touch their former position. Next to them, to their left, we set the nine of the left side of the Second Block of the 81 Set; and next to these again, on their left, the nine of the left side of the Third Block. This Block of twenty-seven now represents Vesper Cubes, which have only one square identical with the Mala cubes of the previous blocks, from which they were taken.

Similarly the Block which is one inch Ana, can be made by taking the nine cubes which come vertically in the middle of each of the Blocks in the first position, and arranging them in a similar manner. Lastly, the Block which lies two inches Ana, can be made by taking the right sides of nine cubes each from each of the three original Blocks, and arranging them so that those in the Second original Block go to the left of those in the First, and those in the Third to their left. In this manner we should obtain three new Blocks, which represent what we can see of the tessaracts, when the direction in which Urna, Moles, Saltus lie in the original Set, is turned W.

The Pluvium Block we can make by taking the front wall of each original Block, and setting each an inch nearer to us, that is -Y. The far sides of these cubes are Moenas of Pluviums. By continuing this treatment of the other walls of the three original Blocks parallel to the front wall, we obtain two other Blocks of tessaracts. The three together are the tessaractic position *a c ō d*, for in all of them Ops lies in the -Y direction, and Dos has been turned W.

The Lar position is more difficult to construct. To put the Lars of the Blocks in their natural position in our space, we must start with the original Mala Blocks, at least three inches above the table. The First Lar Block is made by taking the lowest floors of the three Mala Blocks, and placing them so that that of the Second is below that of the First, and that of the Third below that of the Second. The floor of cubes whose diagonal runs from Urna Lar to Remus Lar, will be at the top of the Block of Lars; and that whose diagonal goes from Cervix Lar to Angusta Lar, will be at the bottom. The next Block of Lars will be made by taking the middle horizontal floors of the three original Blocks, and placing them in a similar succession—the floor from Ostrum Lar to Aer Lar being at the top, that from Cardo Lar to Colus Lar in the middle, and Verbum Lar to Tabula Lar at the bottom. The Third Lar Block is composed of the top floor of the First Block on the top—that is, of Comes Lar to Tyro Lar, of Cortex Lar to Pluma Lar in the middle, and Axis Lar to Portio Lar at the bottom.

# CHAPTER X.
# CYCLICAL PROJECTIONS.

Let us denote the original position of the cube, that wherein Arctos goes Z, Cuspis X, and Dos Y, by the expression,

Z X Y

*a c d*

(1)

If the cube be turned round Cuspis, Dos goes Z, Cuspis remains unchanged, and Arctos goes Y, and we have the position,

Z X Y

*d c a*

where

Z

*d*

means that Dos runs in the negative direction of the Z axis from the point where the axes intersect. We might write

Z

*d*

but it is preferable to write

Z

*d*

. If we next turn the cube round the line, which runs Y, that is, round Arctos, we obtain the position,

Z X Y

*c d a*

(2)

and by means of this double turn we have put *c* and *d* in the places of *a* and *c*. Moreover, we have no negative directions. This position we call simply *c d a*. If from it we turn the cube round *a*, which runs Y, we get

Z

*d*

X

*c*

Y

*a*

, and if, then, we turn it round Dos we get

Z

*d*

X

*a*

Y

*c*

or simply *d a c*. This last is another position in which all the lines are positive, and the projections, instead of lying in different quadrants, will be contained in one.

The arrangement of cubes in *a c d* we know. That in *c d a* is:

Third Floor.

| | | |
|---|---|---|
| Vestis | Oliva | Tyro |
| Scena | Tergum | Aer |
| Saltus | Sypho | Remus |

Second Floor.

| | | |
|---|---|---|
| Tibicen | Mora | Merces |
| Bidens | Pallor | Cortis |
| Moles | Plebs | Hama |

First Floor.

| | | |
|---|---|---|
| Comes | Spicula | Mars |
| Ostrum | Uncus | Ala |
| Urna | Frenum | Sector |

It will be found that learning the cubes in this position gives a great advantage, for thereby the axes of the cube become dissociated with particular directions in space.

The *d a c* position gives the following arrangement:

| | | |
|---|---|---|
| Remus | Aer | Tyro |
| Hama | Cortis | Merces |
| Sector | Ala | Mars |

| | | |
|---|---|---|
| Sypho | Tergum | Oliva |
| Plebs | Pallor | Mora |
| Frenum | Uncus | Spicula |

| | | |
|---|---|---|
| Saltus | Scena | Vestis |
| Moles | Bidens | Tibicen |
| Urna | Ostrum | Comes |

The sides, which touch the vertical plane in the first position, are respectively, in *a c d* Moena, in *c d a* Syce, in *d a c* Alvus.

Take the shape Urna, Ostrum, Moles, Saltus, Scena, Sypho, Remus, Aer, Tyro. This gives in *a c d* the projection: Urna Moena, Ostrum Moena, Moles Moena, Saltus Moena, Scena Moena, Vestis Moena. (If the different positions of the cube are not well known, it is best to have a list of the names before one, but in every case the block should also be built, as well as the names used.) The same shape in the position *c d a* is, of course, expressed in the same words, but it has a different appearance. The front face consists of the Syces of

| Saltus | Sypho | Remus |
|---|---|---|
| Moles | Plebs | Hama |
| Urna | Frenum | Sector |

And taking the shape we find we have Urna, and we know that Ostrum lies behind Urna, and does not come in; next we have Moles, Saltus, and we know that Scena lies behind Saltus and does not come in; lastly, we have Sypho and Remus, and we know that Aer and Tyro are in the Y direction from Remus, and so do not come in. Hence, altogether the projection will consist only of the Syces of Urna, Moles, Saltus, Sypho, and Remus.

Next, taking the position *d a c*, the cubes in the front face have their Alvus sides against the plane, and are:

| Sector | Ala | Mars |
|---|---|---|
| Frenum | Uncus | Spicula |
| Urna | Ostrum | Comes |

And, taking the shape, we find Urna, Ostrum; Moles and Saltus are hidden by Urna, Scena is behind Ostrum, Sypho gives Frenum, Remus gives Sector, Aer gives Ala, and Tyro gives Mars. All these are Alvus sides.

Let us now take the reverse problem, and, given the three cyclical projections, determine the shape. Let the *a c d* projection be the Moenas of Urna, Ostrum, Bidens, Scena, Vestis. Let the *c d a* be the Syces of Urna, Frenum, Plebs, Sypho, and the *d a c* be the Alvus of Urna, Frenum, Uncus, Spicula. Now, from *a c d* we have Urna, Frenum, Sector, Ostrum, Uncus, Ala, Bidens, Pallor, Cortis, Scena, Tergum, Aer, Vestis, Oliva, Tyro. From *c d a* we have Urna, Ostrum, Comes, Frenum, Uncus, Spicula, Plebs, Pallor, Mora, Sypho, Tergum, Oliva. In order to see how these will modify each other, let us consider the *a c d* solution as if it were a set of cubes in the *c d a* arrangement. Here, those that go in the Arctos direction, go away from the plane of projection, and must be represented by the Syce of the cube in contact with the plane. Looking at the *a c d* solution we write down (keeping those together which go away from the plane of projection): Urna and Ostrum, Frenum and Uncus, Sector and Ala, Bidens, Pallor, Cortis, Scena and Vestis, Tergum and Oliva, Aer and Tyro. Here we see that the whole *c d a* face is filled up in the projection, as far as this solution is concerned. But in the *c d a* solution we have only Syces of Urna, Frenum, Plebs, Sypho. These Syces only indicate the presence of a certain number of the cubes stated above as possible from the Moena projection, and those are Urna, Ostrum, Frenum, Uncus, Pallor, Tergum, Oliva. This is the result of a comparison of the Moena projection with the Syce projection. Now, writing these last named as they come in the *d a c* projection, we have Urna, Ostrum, Frenum, Uncus and Pallor and Tergum, Oliva. And of these Ostrum Alvus is wanting in the *d a c* projection as given above. Hence Ostrum will be wanting in the final shape, and we have as the final solution: Urna, Frenum, Uncus, Pallor, Tergum, Oliva.

# CHAPTER XI.
# A TESSARACTIC FIGURE AND ITS PROJECTIONS.

We will now consider a fourth-dimensional shape composed of tessaracts, and the manner in which we can obtain a conception of it. The operation is precisely analogous to that described in chapter VI., by which a plane being could obtain a conception of solid shapes. It is only a little more difficult in that we have to deal with one dimension or direction more, and can only do so symbolically.

We will assume the shape to consist of a certain number of the 81 tessaracts, whose names we have given on p. 168. Let it consist of the thirteen tessaracts: Urna, Moles, Plebs, Frenum, Pallor, Tessera, Cudo, Vitta, Cura, Penates, Polus, Orcus, Lacerta.

Firstly, we will consider what appearances or projections these tessaracts will present to us according as the tessaractic set touches our space with its (*a*) Mala cubes, (*b*) Vesper cubes, (*c*) Pluvium cubes, or (*d*) Lar cubes. Secondly, we will treat the converse question, how the shape can be determined when the projections in each of those views are given.

Let us build up in cubes the four different arrangements of the tessaracts according as they enter our space on their Mala, Vesper, Pluvium or Lar sides. They can only be printed by symbolizing two of the directions. In the following tabulations the directions Y, X will at once be understood. The direction Z (expressed by the wavy line) indicates that the floors of nine, each printed nearer the top of the page, lie above those printed nearer the bottom of it. The direction W is indicated by the dotted line, which shows that the floors of nine lying to the left or right are in the W direction (Ana) or the -W direction (Kata) from those which lie to the right or left. For instance, in the arrangement of the tessaracts, as Malas (Table A) the tessaract Tessara, which is exactly in the middle of the eighty-one tessaracts has

| | | | | |
|---|---|---|---|---|
| Domitor on its right | side | or in the | X | direction. |
| Ocrea on its left | ,, | ,, | -X | ,, |
| Glans away from us | ,, | ,, | Y | ,, |
| Cudo nearer to us | ,, | ,, | -Y | ,, |
| Sacerdos above it | ,, | ,, | Z | ,, |
| Cura below it | ,, | ,, | -Z | ,, |
| Lacerta in the Ana or | | | W | ,, |
| Pallor in the Kata or | | | -W | ,, |

Similarly Cervix lies in the Ana or W direction from Urna, with Thyrsus between them. And to take one more instance, a journey from Saltus to Arcus would be made by travelling Y to Remus, thence -X to Sector, thence Z to Mars, and finally W to Arcus. A line from Saltus to Arcus is therefore a diagonal of the set of 81 tessaracts, because the full length of its side has been traversed in each of the four directions to reach one from the other, *i.e.* Saltus to Remus, Remus to Sector, Sector to Mars, Mars to Arcus.

TABLE A.
Mala presentation of 81 Tessaracts.

Z — W

| Y | | | | Y | | | | Y | | |
|---|---|---|---|---|---|---|---|---|---|---|
| Arcus | Ovis | Portio | | Ara | Vomer | Pluma | | Mars | Merces | Tyro |
| Laurus | Tigris | Segmen | Praeda | Sacerdos | Hydra | Spicula | Mora | Oliva | | |
| Axis | Troja | Aries | Cortex | Mica | Flagellum | Comes | Tibicen | Vestis | | |

X X X

| Y | | | | Y | | | | Y | | |
|---|---|---|---|---|---|---|---|---|---|---|
| Postis | Clipeus | Tabula | | Pilum | Glans | Coins | | Ala | Cortis | Aer |
| *Orcus* | *Lacerta* | Testud | Ocrea | *Tessera* | Domit | Uncu | *Pallor*‡ | Tergu | | |

o or s‡ m

Verbum Luctus Anguis Cardo *Cudo* Malleus Ostrum Bidens‡ Scena

X X

Y Y Y

Telum Nepos Angusta Agmen Lacus Arvus Sector Hama Remus

*Polus* *Penates* Vulcan Crates *Cura* Limen *Frenum‡* *Plebs‡* Sypho

Cervix Securis Vinculum Thyrsus *Vitta* Sceptrum *Urna‡* *Moles‡* Saltus

-Z X

TABLE B.
Vesper presentation of 81 Tessaracts.

W -W

Y Y Y

Porti Plum Tyro Ovis Vom Merc Arc Ara Mar

o a | e e / r s | u s / s

Segmen Hydra Oliva Tigris | Sacerdos Mora Laurus Praeda | Spicula

Aries Flagellum Vestis Troja | Mica Tibicen Axis Cortex | Comes

-X ______ -X ______ -X ______

Y | Y | Y

Tabula Colus Aer | Clipeus Glans Cortis | Postis Pilum Ala

Testudo Domitor Tergum *Lacerta** | *Tessera** *Pallor** *Orcus** Ocrea* | Uncus*

Anguis Malleus Scena Luctus* | *Cudo** Bidens* Verbum† Cardo† | Ostrum†

-X ______ -X ______ -X ______

Y | Y | Y

A A R | N L H | T A S

n r e e a a e g e
g v m p c m l m c
us u u o u a u e t
ta s s s s m n o
r

F
S P re
V Li y Pe C le P C n
ul m p na ur b ol ra u
ca e h tes a* s us te m
n n o * * * s* *

C
Vi Sc S S T U
nc e a ec V M e h rn
ul pt lt ur itt ol r yr a
u ru u is a* es vi su *
m m s * * x s*
*

-Z -X -X

## TABLE C.

Pluvium presentation of 81 Tessaracts.

ZV -W

X X X

M T
M e T S M C i V
a r y p o O o b es
r c r ic r li m ic ti
s e o ul a va e e s
s a s n

V P P S H C M Fl
A o l r a y o ic ag
r m u a c d rt a ell
a e m e er r e u
r a d d a x m
a o

Arcus Ovis Portio Laurus s Tigris Segmen Axis Troja Aries

-Y -Y -Y

X X X

Ala Cortis Aer

Uncus* *Pallor** Tergum

Ostrum† Bidens† Scena

Pilum Glans Colus Ocrea* *Tessera**

Domitor Cardo† *Cudo** Malleus

Postis Clipeus Tabula *Orcus** *Lacerta**

Testudo Verbum† Luctus† Anguis

-Y -Y -Y

X X X

Sector Hama Remus

*Frenum** *Plebs** Sypho

*Urna** *Moles** Saltus

Ag La Ar Cr *Cu* Li Th *Vitt* Sce

m c v at *r*   m yr *a* pt
e u u e *a*   e s * ru
n s s s *   n u m
* s*

S
A *P*
N *P* V C e Vi
T n *e*
e *ol* ul er c n
el g *n*
p *u* c vi u cu
u u *at*
o *s* a x ri lu
m st *es*
s * n † s m
a *
†

-Z -Y -Y -Y

## TABLE D.
Lar presentation of 81 Tessaracts.

ZY -Y

Y Y Y

R
M H
C Se e
M T
A A a
er or ct m
ar yr
la er m
ce tis or u
s o
a
s s

S T *Fr* S
O *P*
pi M U er *en* *Pl* y
li
c or nc *all* g *u* *eb* p
v *or*
ul a us u *m* *s** h
a *
a m * o

S
C Ti V O Bi Sc *U* *M* al
o bi es str d e *rn* *ole* t
m ce ti u e n *a** *s** u
es n s m ns a s

X X X

Y Y Y
A V Pl Pi G C A La A
ra o u lu la ol g ur rv

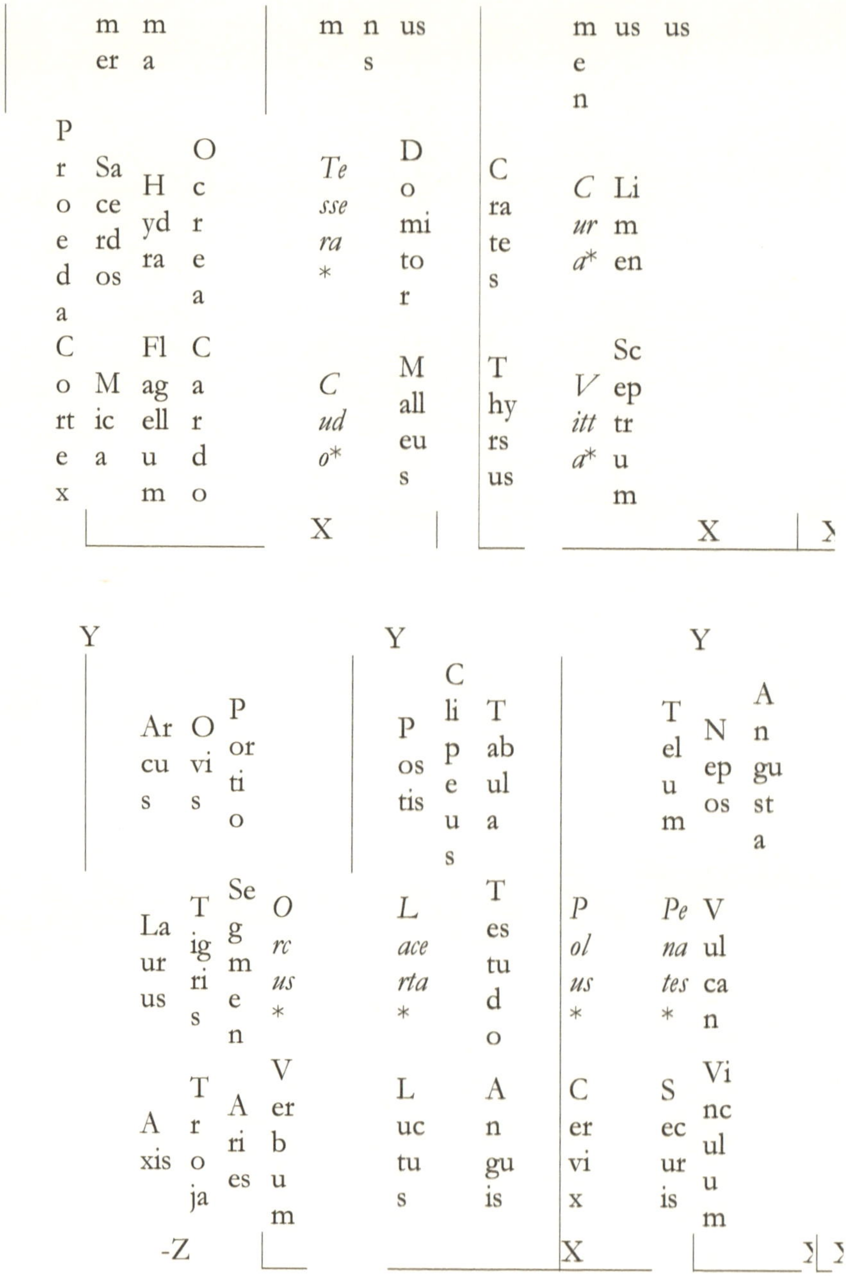


The relation between the four different arrangements shown in the tables A, B, C, and D, will be understood from what has been said in chapter VIII. about a small set of sixteen tessaracts. A glance at the lines, which indicate the directions in each, will show the changes effected by turning the tessaracts from the Mala presentation.

In the Vesper presentation:

The tessaracts—

| | | | | |
|---|---|---|---|---|
| (*e.g.* Urna, Ostrum, Comes), | which ran | Z | still run | Z. |
| (*e.g.* Urna, Frenum, Sector), | „ | Y | „ | Y. |
| (*e.g.* Urna, Moles, Saltus), | „ | X | now run | W. |
| (*e.g.* Urna, Thyrsus, Cervix), | „ | W | „ | -X. |

In the Pluvium presentation:

The tessaracts—

| | | | | |
|---|---|---|---|---|
| (*e.g.* Urna, Ostrum, Comes), | which ran | Z | still run | Z. |
| (*e.g.* Urna, Moles, Saltus), | „ | X | „ | X. |
| (*e.g.* Urna, Frenum, Sector), | „ | Y | now run | W. |
| (*e.g.* Urna, Thyrsus, Cervix), | „ | W | „ | -Y. |

In the Lar presentation:

The tessaracts—

| | | | | |
|---|---|---|---|---|
| (*e.g.* Urna, Moles, Saltus), | which ran | X | still run | X. |
| (*e.g.* Urna, Frenum, Sector), | „ | Y | „ | Y. |
| (*e.g.* Urna, Ostrum, Comes), | „ | Z | now run | W. |
| (*e.g.* Urna, Thyrsus, Cervix), | „ | W | „ | -Z. |

This relation was already treated in chapter IX., but it is well to have it very clear for our present purpose. For it is the apparent change of the relative positions of the tessaracts in each presentation, which enables us to determine any body of them.

In considering the projections, we always suppose ourselves to be situated Ana or W towards the tessaracts, and any movement to be Kata or -W through our space. For instance, in the Mala presentation we have first in our space the Malas of that block of tessaracts, which is the last in the -W direction. Thus, the Mala projection of any given tessaract of the set is that Mala in the extreme -W block, whose place its (the given tessaract's) Mala would occupy, if the tessaractic set moved Kata until the given tessaract reached our space. Or, in other words, if all the tessaracts were transparent except those which constitute the body under consideration, and if a light shone through Four-space from the Ana (W) side to the Kata (-W) side, there would be darkness in each of those Malas, which would be occupied by the Mala of any opaque tessaract, if the tessaractic set moved Kata.

Let us look at the set of 81 tessaracts we have built up in the Mala arrangements, and trace the projections in the extreme -W block of the thirteen of our shape. The latter are printed in italics in Table A, and their projections are marked ‡.

Thus the cube Uncus Mala is the projection of the tessaract Orcus, Pallor Mala of Pallor and Tessera and Tacerta, Bidens Mala of Cudo, Frenum Mala of Frenum and Polus, Plebs Mala of Plebs and Cura and Penates, Moles Mala of Moles and Vitta, Urna Mala of Urna.

Similarly, we can trace the Vesper projections (Table B). Orcus Vesper is the projection of the tessaracts Orcus and Lacerta, Ocrea Vesper of Tessera, Uncus Vesper of Pallor, Cardo Vesper of Cudo, Polus Vesper of Polus and Penates, Crates Vesper of Cura, Frenum Vesper of Frenum and Plebs, Urna Vesper of Urna and Moles, Thyrsus Vesper of Vitta. Next in the Pluvium presentation (Table C) we find that Bidens Pluvium is the projection of the tessaract Pallor, Cudo Pluvium of Cudo and Tessera, Luctus Pluvium of Lacerta, Verbum Pluvium of Orcus, Urna Pluvium of Urna and Frenum, Moles Pluvium of Moles and Plebs, Vitta Pluvium of Vitta and Cura, Securis Pluvium of Penates, Cervix Pluvium of Polus. Lastly, in the Lar presentation (Table D) we observe that Frenum Lar is the projection of Frenum, Plebs Lar of Plebs and Pallor, Moles Lar of Moles, Urna Lar of Urna, Cura Lar of Cura and Tessara, Vitta Lar of Vitta and Cudo, Penates Lar of Penates and Lacerta, Polur Lar of Polus and Orcus.

Secondly, we will treat the converse problem, how to determine the shape when the projections in each presentation are given. Looking back at the list just given above, let us write down in each presentation the projections only.

Mala projections:
Uncus, Pallor, Bidens, Frenum, Plebs, Moles, Urna.
Vesper projections:
Orcus, Ocrea, Uncus, Cardo, Polus, Crates, Frenum, Urna, Thyrsus.
Pluvium projections:
Bidens, Cudo, Luctus, Verbum, Urna, Moles, Vitta, Securis, Cervix.
Lar projections:
Frenum, Plebs, Moles, Urna, Cura, Vitta, Polus, Penates.

Now let us determine the shape indicated by these projections. In now using the same tables we must not notice the italics, as the shape is supposed to be unknown. It is assumed that the reader is building the problem in cubes. From the Mala projections we might infer the presence of all or any of the tessaracts written in the brackets in the following list of the Mala presentation.

(Uncus, Ocrea, Orcus); (Pallor, Tessera, Lacerta);
(Bidens, Cudo, Luctus); (Frenum, Crates, Polus);

(Plebs, Cura, Penates); (Moles, Vitta, Securis);
(Urna, Thyrsus, Cervix).

Let us suppose them all to be present in our shape, and observe what their appearance would be in the Vesper presentation. We mark them all with an asterisk in Table B. In addition to those already marked we must mark (†) Verbum, Cardo, Ostrum, and then we see all the Vesper projections, which would be formed by all the tessaracts possible from the Mala projections. Let us compare these Vesper projections, viz. Orcus, Ocrea, Uncus, Verbum, Cardo, Ostrum, Polus, Crates, Frenum, Cervix, Thyrsus, Urna, with the given Vesper projections. We see at once that Verbum, Ostrum, and Cervix are absent. Therefore, we may conclude that all the tessaracts, which would be implied as possible by their presence, are absent, and of the Mala possibilities may exclude the tessaracts Bidens, Luctus, Securis, and Cervix itself. Thus, of the 21 tessaracts possible in the Mala view, there remain only 17 possible, both in the Mala and Vesper views, viz. Uncus, Ocrea, Orcus, Pallor, Tessera, Lacerta, Cudo, Frenum, Crates, Polus, Plebs, Cura, Penates, Moles, Vitta, Urna, Thyrsus. This we call the Mala-Vesper solution.

Next let us take the Pluvium presentation. We again mark with an asterisk in Table C the possibilities inferred from the Mala-Vesper solution, and take the projections those possibilities would produce. The additional projections are again marked (†). There are twelve Pluvium projections altogether, viz. Bidens, Ostrum, Cudo, Cardo, Luctus, Verbum, Urna, Moles, Vitta, Thyrsus, Securis, Cervix. Again we compare these with the given Pluvium projections, and find three are absent, viz. Ostrum, Cardo, Thyrsus. Hence the tessaracts implied by Ostrum and Cardo and Thyrsus cannot be in our shape, viz. Uncus, Ocrea, Crates, nor Thyrsus itself. Excluding these four from the seventeen possibilities of the Mala-Vesper solution we have left the thirteen tessaracts: Orcus, Pallor, Tessera, Lacerta, Cudo, Frenum, Polus, Plebs, Cura, Penates, Moles, Vitta, Urna. This we call the Mala-Vesper-Pluvium solution.

Lastly, we have to consider whether these thirteen tessaracts are consistent with the given Lar projections. We mark them again on Table D with an asterisk, and we find that the projections are exactly those given, viz. Frenum, Plebs, Moles, Urna, Cura, Vitta, Polus, Penates. Therefore, we have not to exclude any of the thirteen, and can infer that they constitute the shape, which produces the four different given views or projections.

In fine, any shape in space consists of the possibilities common to the projections of its parts upon the boundaries of that space, whatever be the number of its dimensions. Hence the simple rule for the determination of the shape would be to write down all the possibilities of the sets of projections, and then cancel all those possibilities which are not common to all. But the process adopted above is much preferable, as through it we may realize the gradual delimitation of the shape view by view. For once more we must remind ourselves that our great object is, not to arrive at results by symbolical operations, but to realize those results piece by piece through realized processes.

—

# APPENDICES.

## APPENDIX A.

This set of 100 names is useful for studying Plane Space, and forms a square 10 × 10.

| | | | | | | | | | |
|---|---|---|---|---|---|---|---|---|---|
| Aiōn | Bios | Hupar | Neas | Kairos | Enos | Thlipsis | Cheimas | Theion | Epei |
| Itea | Hagios | Phaino | Geras | Tholos | Ergon | Pachūs | Kiōn | Eris | Cleos |
| Loma | Etēs | Trochos | Klazo | Lutron | Hēdūs | Ischūs | Paigma | Hedna | Demas |
| Numphe | Bathus | Pauo | Euthu | Holos | Para | Thuos | Karē | Pylē | Spareis |
| Ania | Eōn | Seranx | Mesoi | Dramo | Thallos | Aktē | Ozo | Onos | Magos |
| Notos | Mēnis | Lampas | Ornis | Thama | Eni | Pholis | Mala | Strizo | Rudon |
| Labo | Helor | Rupa | Rabdos | Doru | Epos | Theos | Idris | Ēdē | Hepo |
| Sophos | Ichor | Kaneōn | Ephthra | Oxis | Lukē | Blue | Helos | Peri | Thelus |
| Eunis | Limos | Keedo | Igde | Matē | Lukos | Pteris | Holmos | Oulo | Dokos |
| Aeido | Ias | Assa | Muzo | Hippeus | Eōs | Atē | Akme | Ōrē | Gua |

## APPENDIX B.

The following list of names is used to denote cubic spaces. It makes a cubic block of six floors, the highest being the sixth.

| | | | | | | |
|---|---|---|---|---|---|---|
| | Fons | Plect rum | Vulnu s | Are na | Me nsa | Ter minu s |
| I S | Testa | Plau sus | Uva | Colli s | Co ma | Neb ula |
| l i | Copi a | Cor nu | Solu m | Mun us | Rix um | Vitru m |
| o | Ars | Ferv or | Thym a | Col ubra | Seg es | Cor |
| o t | Lupu s | Clas sis | Modu s | Fla mm a | Me ns | Incol a |
| kr | Thal amus | Hast a | Cala mus | Crin is | Aur iga | Vallu m |

| | | | | | | | |
|---|---|---|---|---|---|---|---|
| *F i f* | *F l o* | Linte um | Pinn is | Pup pis | Nupt ia | Aegi s | Cith ara |
| | | Triu | Curr | Lux | Port | Latu | Funi |

th or.

mphus is us s s

Regnum Fascis Bellum Capellus Arbor Custos
Sagitta Puer Stella Saxum Humor Pontus
Nomen Imago Lapsus Quercus Mundus Proelium
Palaestra Nuncius Bos Pharetra Pumex Tibia

Lignum Focus Ornus Lucrum Alea Vox
Caterva Facies Onus Silva Gelu Flumen
Tellus Sol Os Arma Brachium Jaculum
Merum Signum Umbra Tempus Corona Socius
Moena Opus Honor Campus Rivus Imber
Victor Equus Miles Cursus Lyra Tunica

Fourth Floor.

Haedus Taberna Turris Nox Domus Vinum
Pruinus Chorus Luna Flos Lucus Agna
Fulmen Hiems Ver Carina Arator Pratum
Oculus Ignis Aether Cohors Penna Labor
Aes Pectus Pelagus Notus Fretum Gradus
Princeps Dux Ventus Navis Finis Robur

Third Floor.

Vultus Hostis Figura Ales Coelum Aura
Humerus Augur Ludus Clamor Galea Pes
Civis Ferrum Pugna Res Carmen Nubes

Second Floor.

Litus Unda Rex Templum Ripa Amnis
Pannus Ulmus Sedes Columba Aequor Dama
Dexter Urbs Gens Monstrum Pecus Mons

First Floor.

Nemus Sidus Vertex Nix Grando Arx
Venator Cerva Aper Plaga Hedera Frons
Mem Aqua Cap Castr Litu Tu

| brum | | ut | um | us | ba |
|---|---|---|---|---|---|
| Fluct us | Rus | Rati s | Amp hora | Pars | Di es |
| Turb a | Ager | Tra bs | Myrt us | Fibr a | Na uta |
| Decu s | Pulvi s | Met a | Rota | Pal ma | Te rra |

# APPENDIX C.

The following names are used for a set of 256 Tessaracts.

## FOURTH BLOCK.

*Fourth Floor.*

| Dolium | Caballus | Python | Circaea |
|---|---|---|---|
| Cussis | Pulsus | Drachma | Cordax |
| Porrum | Consul | Diota | Dyka |
| Columen | Ravis | Corbis | Rapina |

*Third Floor.*

| Alexis | Planta | Corymbus | Lectrum |
|---|---|---|---|
| Aestus | Labellum | Calathus | Nux |
| Septum | Sepes | Turtur | Ordo |
| Morsus | Aestas | Capella | Rheda |

*Second Floor.*

| Corydon | Jugum | Tornus | Labrum |
|---|---|---|---|
| Lac | Hibiscus | Donum | Caltha |
| Senex | Palus | Salix | Cespes |
| Amictus | Gurges | Otium | Pomum |

*First Floor.*

| Odor | Aprum | Pignus | Messor |
|---|---|---|---|
| Color | Casa | Cera | Papaver |
| Spes | Lapis | Apis | Afrus |
| Vitula | Clavis | Fagus | Cornix |

## THIRD BLOCK.

*Fourth Floor.*

| Charta | Cures | Quaestor | Cliens |
|---|---|---|---|
| Frux | Pyra | Lena | Procella |
| Hera | Esca | Secta | Rugæ |
| Eurus | Gloria | Socer | Sequela |

*Third Floor.*

| Arche | Agger | Cumulus | Cassis |
|---|---|---|---|
| Arcus | Ovis | Portio | Mimus |
| Laurus | Tigris | Segmen | Obolus |
| Axis | Troja | Aries | Fuga |

*Second Floor.*

| Ruina | Culmen | Fenestra | Aedes |
|---|---|---|---|
| Postis | Clipeus | Tabula | Lingua |
| Orcus | Lacerta | Testudo | Scala |
| Verbum | Luctus | Anguis | Dolus |

*First Floor.*

| Additus | Salus | Clades | Rana |
|---|---|---|---|
| Telum | Nepos | Angusta | Mucro |
| Polus | Penates | Vulcan | Ira |
| Cervix | Securis | Vinculum | Furor |

## SECOND BLOCK.

*Fourth Floor.*

| Actus | Spadix | Sicera | Anser |
|---|---|---|---|
| Auspex | Praetor | Atta | Sonus |
| Fulgor | Ardea | Prex | Aevum |
| Spina | Birrus | Acerra | Ramus |

*Third Floor.*

| Machina | Lex | Omen | Artus |
|---|---|---|---|
| Ara | Vomer | Pluma | Odium |
| Proeda | Sacerdos | Hydra | Luxus |
| Cortex | Mica | Flagellum | Mas |

*Second Floor.*

| Ardor | Rupes | Pallas | Arista |
|---|---|---|---|
| Pilum | Glans | Colus | Pellis |
| Ocrea | Tessara | Domitor | Fera |

## FIRST BLOCK.

*Fourth Floor.*

| Horreum | Fumus | Hircus | Erisma |
|---|---|---|---|
| Anulus | Pluor | Acies | Naxos |
| Etna | Gemma | Alpis | Arbiter |
| Alauda | Furca | Gena | Alnus |

*Third Floor.*

| Fax | Venenum | Syrma | Ursa |
|---|---|---|---|
| Mars | Merces | Tyro | Fama |
| Spicula | Mora | Oliva | Conjux |
| Comes | Tibicen | Vestis | Plenum |

*Second Floor.*

| Rostrum | Armiger | Premium | Tribus |
|---|---|---|---|
| Ala | Cortis | Aer | Fragor |
| Uncus | Pallor | Tergum | Reus |

| Cardo | Cudo | Malleus | Thorax | Ostrum | Bidens | Scena | Torus |
|---|---|---|---|---|---|---|---|
| | *First Floor.* | | | | *First Floor.* | | |
| Regina | Canis | Marmor | Tectum | Pardus | Rubor | Nurus | Hospes |
| Agmen | Lacus | Arvus | Rumor | Sector | Hama | Remus | Fortuna |
| Crates | Cura | Limen | Vita | Frenum | Plebs | Sypho | Myrrha |
| Thyrsus | Vitta | Sceptrum | Pax | Urna | Moles | Saltus | Acus |

## APPENDIX D.

The following list gives the colours, and the various uses for them. They have already been used in the foregoing pages to distinguish the various regions of the Tessaract, and the different individual cubes or Tessaracts in a block. The other use suggested in the last column of the list has not been discussed; but it is believed that it may afford great aid to the mind in amassing, handling, and retaining the quantities of formulae requisite in scientific training and work.

| *Colour.* | *Region of Tessaract.* | *Tessaract in 81 Set.* | *Symbol.* |
|---|---|---|---|
| Black | Syce | Plebs | 0 |
| White | Mel | Mora | 1 |
| Vermilion | Alvus | Uncus | 2 |
| Orange | Cuspis | Moles | 3 |
| Light-yellow | Murex | Cortis | 4 |
| Bright-green | Lappa | Penates | 5 |
| Bright-blue | Iter | Oliva | 6 |
| Light-grey | Lares | Tigris | 7 |
| Indian-red | Crux | Orcus | 8 |
| Yellow-ochre | Sal | Testudo | 9 |
| Buff | Cista | Sector | + (plus) |
| Wood | Tessaract | Tessara | - (minus) |
| Brown-green | Tholus | Troja | ± (plus or minus) |
| Sage-green | Margo | Lacerta | × (multiplied by) |
| Reddish | Callis | Tibicen | ÷ (divided by) |
| Chocolate | Velum | Sacerdos | = (equal to) |
| French-grey | Far | Scena | ≠ (not equal to) |
| Brown | Arctos | Ostrum | > (greater than) |
| Dark-slate | Daps | Aer | < (less than) |
| Dun | Portica | Clipeus | : } signs of proportion |
| Orange-vermilion | Talus | Portio | :: |

| Stone | Ops | Thyrsus | · (decimal point) |
|---|---|---|---|
| Quaker-green | Felis | Axis | ∟ (factorial) |
| Leaden | Semita | Merces | ‖ (parallel) |
| Dull-green | Mappa | Vulcan | ∦ (not parallel) |
| Indigo | Lixa | Postis | $\pi/2$ (90°) (at right angles) |
| Dull-blue | Pagus | Verbum | log. base 10 |
| Dark-purple | Mensura | Nepos | sin. (sine) |
| Pale-pink | Vena | Tabula | cos. (cosine) |
| Dark-blue | Moena | Bidens | tan. (tangent) |
| Earthen | Mugil | Angusta | ∞ (infinity) |
| Blue | Dos | Frenum | a |
| Terracotta | Crus | Remus | b |
| Oak | Idus | Domitor | c |
| Yellow | Pagina | Cardo | d |
| Green | Bucina | Ala | e |
| Rose | Olla | Limen | f |
| Emerald | Orsa | Ara | g |
| Red | Olus | Mars | h |
| Sea-green | Libera | Pluma | i |
| Salmon | Tela | Glans | j |
| Pale-yellow | Livor | Ovis | k |
| Purple-brown | Opex | Polus | l |
| Deep-crimson | Camoena | Pilum | m |
| Blue-green | Proes | Tergum | n |
| Light-brown | Lua | Crates | o |
| Deep-blue | Lama | Tyro | p |
| Brick-red | Lar | Cura | q |
| Magenta | Offex | Arvus | r |
| Green-grey | Cadus | Hama | s |
| Light-red | Croeta | Praeda | t |
| Azure | Lotus | Vitta | u |
| Pale-green | Vesper | Ocrea | v |
| Blue-tint | Panax | Telum | w |
| Yellow-green | Pactum | Malleus | x |
| Deep-green | Mango | Vomer | y |
| Light-green | Lis | Agmen | z |
| Light-blue | Ilex | Comes | α |
| Crimson | Bolus | Sypho | β |
| Ochre | Limbus | Mica | γ |
| Purple | Solia | Arcus | δ |
| Leaf-green | Luca | Securis | ε |
| Turquoise | Ancilla | Vinculum | ζ |
| Dark-grey | Orca | Colus | η |
| Fawn | Nugæ | Saltus | θ |
| Smoke | Limus | Sceptrum | ι |
| Light-buff | Mala | Pallor | κ |
| Dull-purple | Sors | Vestis | λ |
| Rich-red | Lucta | Cortex | μ |

| | | | |
|---|---|---|---|
| Green-blue | Pator | Flagellum | ν |
| Burnt-sienna | Silex | Luctus | ξ |
| Sea-blue | Lorica | Lacus | ο |
| Peacock-blue | Passer | Aries | π |
| Deep-brown | Meatus | Hydra | ϱ |
| Dark-pink | Onager | Anguis | σ |
| Dark | Lensa | Laurus | τ |
| Dark-stone | Pluvium | Cudo | υ |
| Silver | Spira | Cervix | φ |
| Gold | Corvus | Urna | χ |
| Deep-yellow | Via | Spicula | ψ |
| Dark-green | Calor | Segmen | ω |

## APPENDIX E.

### A THEOREM IN FOUR-SPACE.

If a pyramid on a triangular base be cut by a plane which passes through the three sides of the pyramid in such manner that the sides of the sectional triangle are not parallel to the corresponding sides of the triangle of the base; then the sides of these two triangles, if produced in pairs, will meet in three points which are in a straight line, namely, the line of intersection of the sectional plane and the plane of the base.

Let A B C D be a pyramid on a triangular base A B C, and let a b c be a section such that A B, B C, A C, are respectively not parallel to a b, b c, a c. It must be understood that a is a point on A D, b is a point on B D, and c a point on C D. Let, A B and a b, produced, meet in m. B C and b c, produced, meet in n; and A C and a c, produced, meet in o. These three points, m, n, o, are in the line of intersection of the two planes A B C and a b c.

Now, let the line a b be projected on to the plane of the base, by drawing lines from a and b at right angles to the base, and meeting it in a' b'; the line a' b', produced, will meet A B produced in m. If the lines b c and a c be projected in the same way on to the base, to the points b' c' and a' c'; then B C and b' c' produced, will meet in n, and A C and a' c' produced, will meet in o. The two triangles A B C and a' b' c' are such, that the lines joining A to a', B to b', and C to c', will, if produced, meet in a point, namely, the point on the base A B C which is the projection of D. Any two triangles which fulfil this condition are the possible base and projection of the section of a pyramid; therefore the sides of such triangles, if produced in pairs, will meet (if they are not parallel) in three points which lie in one straight line.

A four-dimensional pyramid may be defined as a figure bounded by a polyhedron of any number of sides, and the same number of pyramids whose bases are the sides of the polyhedron, and whose apices meet in a point not in the space of the base.

If a four-dimensional pyramid on a tetrahedral base be cut by a space which passes through the four sides of the pyramid in such a way that the sides of the sectional figure be not parallel to the sides of the base; then the sides of these two tetrahedra, if produced in pairs, will meet in lines which all lie in one plane, namely, the plane of intersection of the space of the base and the space of the section.

If now the sectional tetrahedron be projected on to the base (by drawing lines from each point of the section to the base at right angles to it), there will be two tetrahedra fulfilling the condition that the line joining the angles of the one to the angles of the other will, if produced, meet in a point, which point is the projection of the apex of the four-dimensional pyramid.

Any two tetrahedra which fulfil this condition, are the possible base and projection of a section of a four-dimensional pyramid. Therefore, in any two such tetrahedra, where the sides of the one are not parallel to the sides of the other, the sides, if produced in pairs (one side of the one with one side of the other), will meet in four straight lines which are all in one plane.

## APPENDIX F.

### EXERCISES ON SHAPES OF THREE DIMENSIONS.

The names used are those given in Appendix B.

Find the shapes from the following projections:

1. Syce projections: Ratis, Caput, Castrum, Plagua.
Alvus projections: Merum, Oculus, Fulmen, Pruinus.
Moena projections: Miles, Ventus, Navis.

2. Syce: Dies, Tuba, Lituus, Frons.
Alvus: Sagitta, Regnum, Tellus, Fulmen, Pruinus.

Moena: Tibia, Tunica, Robur, Finis.

3. Syce: Nemus, Sidus, Vertex, Nix, Cerva.

Alvus: Lignum, Haedus, Vultus, Nemus, Humerus.

Moena: Dexter, Princeps, Equus, Dux, Urbs, Pullis, Gens, Monstrum, Miles.

4. Syce: Amphora, Castrum, Myrtus, Rota, Palma, Meta, Trabs, Ratis.

Alvus: Dexter, Princeps, Moena, Aes, Merum, Oculus, Littus, Civis, Fulmen.

Moena: Gens, Ventus, Navis, Finis, Monstrum, Cursus.

5. Syce: Castrum, Plagua, Nix, Vertex, Aper, Caput, Cerva, Venator.

Alvus: Triumphus, Tellus, Caterva, Lignum, Haedus, Pruinus, Fulmen, Civis, Humerus, Vultus.

Moena: Pharetra, Cursus, Miles, Equus, Dux, Navis, Monstrum, Gens, Urbs, Dexter.

## ANSWERS.

The shapes are:

1. Umbra, Aether, Ver, Carina, Flos.
2. Pontus, Custos, Jaculum, Pratum, Arator, Agna.
3. Focus, Omus, Haedus, Tabema, Vultus, Hostis, Figura, Ales, Sidus, Augur.
4. Tempus, Campus, Finis, Navis, Ventus, Pelagus, Notus, Cohors, Aether, Carina, Res, Templum, Rex, Gens, Monstrum.
5. Portus, Arma, Sylva, Lucrum, Ornus, Onus, Os, Facies, Chorus, Carina, Flos, Nox, Ales, Clamor, Res, Pugna, Ludus, Figura, Augur, Humerus.

## FURTHER EXERCISES IN SHAPES OF THREE DIMENSIONS.

The Names used are those given in Appendix C; and this set of exercises forms a preparation for their use in space of four dimensions. All are in the 27 Block (Urna to Syrma).

1. Syce: Moles, Frenum, Plebs, Sypho.

Alvus: Urna, Frenum, Uncus, Spicula, Comes.

Moena: Moles, Bidens, Tibicen, Comes, Saltus.

2. Syce: Urna, Moles, Plebs, Hama, Remus.

Alvus: Urna, Frenum, Sector, Ala, Mars.

Moena: Urna, Moles, Saltus, Bidens, Tibicen.

3. Syce: Moles, Plebs, Hama, Remus.
Alvus: Uma, Ostrum, Comes, Spicula, Frenum, Sector.
Moena: Moles, Saltus, Bidens, Tibicen.
4. Syce: Frenum, Plebs, Sypho, Moles, Hama.
Alvus: Urna, Frenum, Uncus, Sector, Spicula.
Moena: Urna, Moles, Saltus, Scena, Vestis.
5. Syce: Urna, Moles, Plebs, Hama, Remus, Sector.
Alvus: Urna, Frenum, Sector, Uncus, Spicula, Comes, Mars.
Moena: Urna, Moles, Saltus, Bidens, Tibicen, Comes.
6. Syce: Uma, Moles, Saltus, Sypho, Remus, Hama, Sector.
Alvus: Comes, Ostrum, Uncus, Spicula, Mars, Ala, Sector.
Moena: Urna, Moles, Saltus, Scena, Vestis, Tibicen, Comes, Ostrum.
7. Syce: Sypho, Saltus, Moles, Urna, Frenum, Sector.
Alvus: Urna, Frenum, Uncus, Spicula, Mars.
Moena: Saltus, Moles, Urna, Ostrum, Comes.
8. Syce: Moles, Plebs, Hama, Sector.
Alvus: Ostrum, Frenum, Uncus, Spicula, Mars, Ala.
Moena: Moles, Bidens, Tibicen, Ostrum.
9. Syce: Moles, Saltus, Sypho, Plebs, Frenum, Sector.
Alvus: Ostrum, Comes, Spicula, Mars, Ala.
Moena: Ostrum, Comes, Tibicen, Bidens, Scena, Vestis.
10. Syce: Urna, Moles, Saltus, Sypho, Remus, Sector, Frenum.
Alvus: Urna, Ostrum, Comes, Spicula, Mars, Ala, Sector.
Moena: Urna, Ostrum, Comes, Tibicen, Vestis, Scena, Saltus.
11. Syce: Frenum, Plebs, Sypho, Hama.
Alvus: Frenum, Sector, Ala, Mars, Spicula.
Moena: Urna, Moles, Saltus, Bidens, Tibicen.

## Answers.

The shapes are:

1. Moles, Plebs, Sypho, Pallor, Mora, Tibicen, Spicula.
2. Urna, Moles, Plebs, Hama, Cortis, Merces, Remus.
3. Moles, Bidens, Tibicen, Mora, Plebs, Hama, Remus.
4. Frenum, Plebs, Sypho, Tergum, Oliva, Moles, Hama.
5. Urna, Moles, Plebs, Hama, Remus, Pallor, Mora, Tibicen, Mars, Merces, Comes, Sector.
6. Ostrum, Comes, Tibicen, Vestis, Scena, Tergum, Oliva, Tyro, Aer, Remus, Hama, Sector, Merces, Mars, Ala.
7. Sypho, Saltus, Moles, Urna, Frenum, Uncus, Spicula, Mars.
8. Plebs, Pallor, Mora, Bidens, Merces, Cortis, Ala.

9. Bidens, Tibicen, Vestis, Scena, Oliva, Mora, Spicula, Mars, Ala.
10. Urna, Ostrum, Comes, Spicula, Mars, Tibicen, Vestis, Oliva, Tyro, Aer, Remus, Sector, Ala, Saltus, Scena.
11. Frenum, Plebs, Sypho, Hama, Cortis, Merces, Mora.

APPENDIX G.

## EXERCISES ON SHAPES OF FOUR DIMENSIONS.

The Names used are those given in Appendix C. The first six exercises are in the 81 Set, and the rest in the 256 Set.

1. Mala projection: Urna, Moles, Plebs, Pallor, Cortis, Merces.
Lar projection: Urna, Moles, Plebs, Cura, Penates, Nepos.
Pluvium projection: Urna, Moles, Vitta, Cudo, Luctus, Troja.
Vesper projection: Urna, Frenum, Crates, Ocrea, Orcus, Postis, Arcus.

2. Mala: Urna, Frenum, Uncus, Pallor, Cortis, Aer.
Lar: Urna, Frenum, Crates, Cura, Lacus, Arvus, Angusta.
Pluvium: Urna, Thyrsus, Cardo, Cudo, Malleus, Anguis.
Vesper: Urna, Frenum, Crates, Ocrea, Pilum, Postis.

3. Mala: Comes, Tibicen, Mora, Pallor.
Lar: Urna, Moles, Vitta, Cura, Penates.
Pluvium: Comes, Tibicen, Mica, Troja, Luctus.
Vesper: Comes, Cortex, Praeda, Laurus, Orcus.

4. Mala: Vestis, Oliva, Tyro.
Lar: Saltus, Sypho, Remus, Arvus, Angusta.
Pluvium: Vestis, Flagellum, Aries.
Vesper: Comes, Spicula, Mars, Ara, Arcus.

5. Mala: Mars, Merces, Tyro, Aer, Tergum, Pallor, Plebs.
Lar: Sector, Hama, Lacus, Nepos, Angusta, Vulcan, Penates.
Pluvium: Comes, Tibicen, Mica, Troja, Aries, Anguis, Luctus, Securis.
Vesper: Mars, Ara, Arcus, Postis, Orcus, Polus.

6. Mala: Pallor, Mora, Oliva, Tyro, Merces, Mars, Spicula, Comes, Tibicen, Vestis.
Lar: Plebs, Cura, Penates, Vulcan, Angusta, Nepos, Telum, Polus, Cervix, Securis, Vinculum.
Pluvium: Bidens, Cudo, Luctus, Troja, Axis, Aries.

Vesper: Uncus, Ocrea, Orcus, Laurus, Arcus, Axis.

7. Mala: Hospes, Tribus, Fragor, Aer, Tyro, Mora, Oliva.

Lar: Hospes, Tectum, Rumor, Arvus, Angusta, Cera, Apis, Lapis.

Pluvium: Acus, Torus, Malleus, Flagellum, Thorax, Aries, Aestas, Capella.

Vesper: Pardus, Rostrum, Ardor, Pilum, Ara, Arcus, Aestus, Septum.

8. Mala: Pallor, Tergum, Aer, Tyro, Cortis, Syrma, Ursa, Fama, Naxos, Erisma.

Lar: Plebs, Cura, Limen, Vulcan, Angusta, Nepos, Cera, Papaver, Pignus, Messor.

Pluvium: Bidens, Cudo, Malleus, Anguis, Aries, Luctus, Capella, Rheda, Rapina.

Vesper: Uncus, Ocrea, Orcus, Postis, Arcus, Aestus, Cussis, Dolium, Alexis.

9. Mala: Fama, Conjux, Reus, Torus, Acus, Myrrha, Sypho, Plebs, Pallor, Mora, Oliva, Alpis, Acies, Hircus.

Lar: Missale, Fortuna, Vita, Pax, Furor, Ira, Vulcan, Penates, Lapis, Apis, Cera, Pignus.

Pluvium: Torus, Plenum, Pax, Thorax, Dolus, Furor, Vinculum, Securis, Clavis, Gurges, Aestas, Capella, Corbis.

Vesper: Uncus, Spicula, Mars, Ocrea, Cardo, Thyrsus, Cervix, Verbum, Orcus, Polus, Spes, Senex, Septum, Porrum, Cussis, Dolium.

## ANSWERS.

The shapes are:

1. Urna, Moles, Plebs, Cura, Tessara, Lacerta, Clipeus, Ovis.
2. Urna, Frenum, Crates, Ocrea, Tessara, Glans, Colus, Tabula.
3. Comes, Tibicen, Mica, Sacerdos, Tigris, Lacerta.
4. Vestis, Oliva, Tyro, Pluma, Portio.
5. Mars, Merces, Vomer, Ovis, Portio, Tabula, Testudo, Lacerta, Penates.
6. Pallor, Tessara, Lacerta, Tigris, Segmen, Portio, Ovis, Arcus, Laurus, Axis, Troja, Aries.
7. Hospes, Tribus, Arista, Pellis, Colus, Pluma, Portio, Calathus, Turtur, Sepes.
8. Pallor, Tessara, Domitor, Testudo, Tabula, Clipeus, Portio, Calathus, Nux, Lectrum, Corymbus, Circaea, Cordax.

9. Fama, Conjux, Reus, Fera, Thorax, Pax, Furor, Dolus, Scala, Ira, Vulcan, Penates, Lapis, Palus, Sepes, Turtur, Diota, Drachma, Python.

-

## APPENDIX H.

### SECTIONS OF CUBE AND TESSARACT.

There are three kinds of sections of a cube.

1. The sectional plane, which is in all cases supposed to be infinite, can be taken parallel to two of the opposite faces of the cube; that is, parallel to two of the lines meeting in Corvus, and cutting the third.

2. The sectional plane can be taken parallel to one of the lines meeting in Corvus and cutting the other two, or one or both of them produced.

3. The sectional plane can be taken cutting all three lines, or any or all of them produced.

Take the first case, and suppose the plane cuts Dos half-way between Corvus and Cista. Since it does not cut Arctos or Cuspis, or either of them produced, it will cut Via, Iter, and Bolus at the middle point of each; and the figure, determined by the intersection of the Plane and Mala, is a square. If the length of the edge of the cube be taken as the unit, this figure may be expressed thus:

Z
0

.

X
0

.

Y
$\frac{1}{2}$

showing that the Z and X lines from Corvus are not cut at all, and that the Y line is cut at half-a-unit from Corvus.

Sections taken

Z
0
.
X
0
.
Y
1/4

and

Z
0
.
X
0
.
Y
1

would also be squares.

Take the second case.

Let the plane cut Cuspis and Dos, each at half-a-unit from Corvus, and not cut Arctos or Arctos produced; it will also cut through the middle points of Via and Callis. The figure produced, is a rectangle which has two sides of one unit, and the other two are each the diagonal of a half-unit squared.

If the plane cuts Cuspis and Dos, each at one unit from Corvus, and is parallel to Arctos, the figure will be a rectangle which has two sides of one unit in length; and the other two the diagonal of one unit squared.

If the plane passes through Mala, cutting Dos produced and Cuspis produced, each at one-and-a-half unit from Corvus, and is parallel to Arctos, the figure will be a parallelogram like the one obtained by the section

Z
0
.
X
/
1 2
.
Y
/
1 2

.

This set of figures will be expressed

Z
0
.
X
/
1 2
.
Y
/
1 2

Z

0

.

X
1

.

Y
1

Z
0

.

X
1⁄
2

.

Y
1⁄
2

It will be seen that these sections are parallel to each other; and that in each figure Cuspis and Dos are cut at equal distances from Corvus.

We may express the whole set thus:—

Z
O

.

X

I

.

Y
I

it being understood that where Roman figures are used, the numbers do not refer to the length of unit cut off any given line from Corvus, but to the proportion between the lengths. Thus

Z
O

.

X
I

.

Y
II

means that Arctos is not cut at all, and that Cuspis and Dos are cut, Dos being cut twice as far from Corvus as is Cuspis.

These figures will also be rectangles.

Take the third case.

Suppose Arctos, Cuspis, and Dos are each cut half-way. This figure is an equilateral triangle, whose sides are the diagonal of a half-unit squared. The figure

Z
1

.

X
1

.

Y
1

is also an equilateral triangle, and the figure

Z
1/2

.

X
1/2

.

Y
1/2

is an equilateral hexagon.

It is easy for us to see what these shapes are, and also, what the figures of any other set would be, as

Z
I

.

X
II

.

Y
II

or

Z
I

.

X
II

.

Y
III

but we must learn them as a two-dimensional being would, so that we may see how to learn the three-dimensional sections of a tessaract.

It is evident that the resulting figures are the same whether we fix the cube, and then turn the sectional plane to the required position, or whether we fix the sectional plane, and then turn the cube. Thus, in the first case we might have fixed the plane, and then so placed the cube that one plane side coincided with the sectional plane, and then have drawn the cube half-way through, in a direction at right angles to the plane, when we should have seen the square first mentioned. In the second case (

Z
O

.

X
I

.

Y
I

)

we might have put the cube with Arctos coinciding with the plane and with Cuspis and Dos equally inclined to it, and then have drawn the cube through the plane at right angles to it until the lines (Cuspis and Dos) were cut at the required distances from Corvus. In the third case we might have put the cube with only Corvus coinciding with the plane and with Cuspis, Dos, and Arctos equally inclined to it (for any of the shapes in the set

Z
I

X
I

Y
I

) and then have drawn it through as before. The resulting figures are exactly the same as those we got before; but this way is the best to use, as it would probably be easier for a two-dimensional being to think of a cube passing through his space than to imagine his whole space turned round, with regard to the cube.

We have already seen (p. 117) how a two-dimensional being would observe the sections of a cube when it is put with one plane side coinciding with his space, and is then drawn partly through.

Now, suppose the cube put with the line Arctos coinciding with his space, and the lines Cuspis and Dos equally inclined to it. At first he would only see Arctos. If the cube were moved until Dos and Cuspis were each cut half-way, Arctos still being parallel to the plane, Arctos would disappear at once; and to find out what he would see he would have to take the square sections of the cube, and find on each of them what lines are given by the new set of sections. Thus he would take Moena itself, which may be regarded as the first section of the square set. One point of the figure would be the middle point of Cuspis, and since the sectional plane is parallel to Arctos, the line of intersection of Moena with the sectional plane will be parallel to Arctos. The required line then cuts Cuspis half-way, and is parallel to Arctos, therefore it cuts Callis half-way.

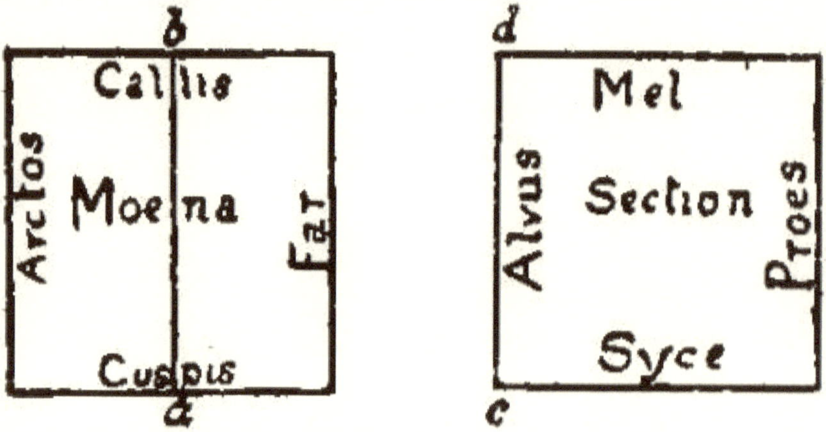


Fig. 21.

Next he would take the square section half-way between Moena and Murex. He knows that the line Alvus of this section is parallel to Arctos, and that the point Dos at one of its ends is half-way between Corvus and Cista, so that this line itself is the one he wants (because the sectional plane cuts Dos half-way between Corvus and Cista, and is parallel to Arctos). In Fig. 21 the two lines thus found are shown. a b is the line in Moena, and c d the line in the section. He must now find out how far apart they are. He knows that from the middle point of Cuspis to Corvus is half-a-unit, and from the middle point of Dos to Corvus is half-a-unit, and Cuspis and Dos are at right angles to each other; therefore from the middle point of Cuspis to the middle point of Dos is the diagonal of a square whose sides are half-a-unit in length. This diagonal may be written d (√½)2. He would also see that from the middle point of Callis to the middle point of Via is the same length; therefore the figure is a parallelogram, having two of its sides, each one unit in length, and the other two each d (√½)2.

He could also see that the angles are right, because the lines a c and b d are made up of the X and Y directions, and the other two, a b and d, are purely Z, and since they have no tendency in common, they are at right angles to each other.

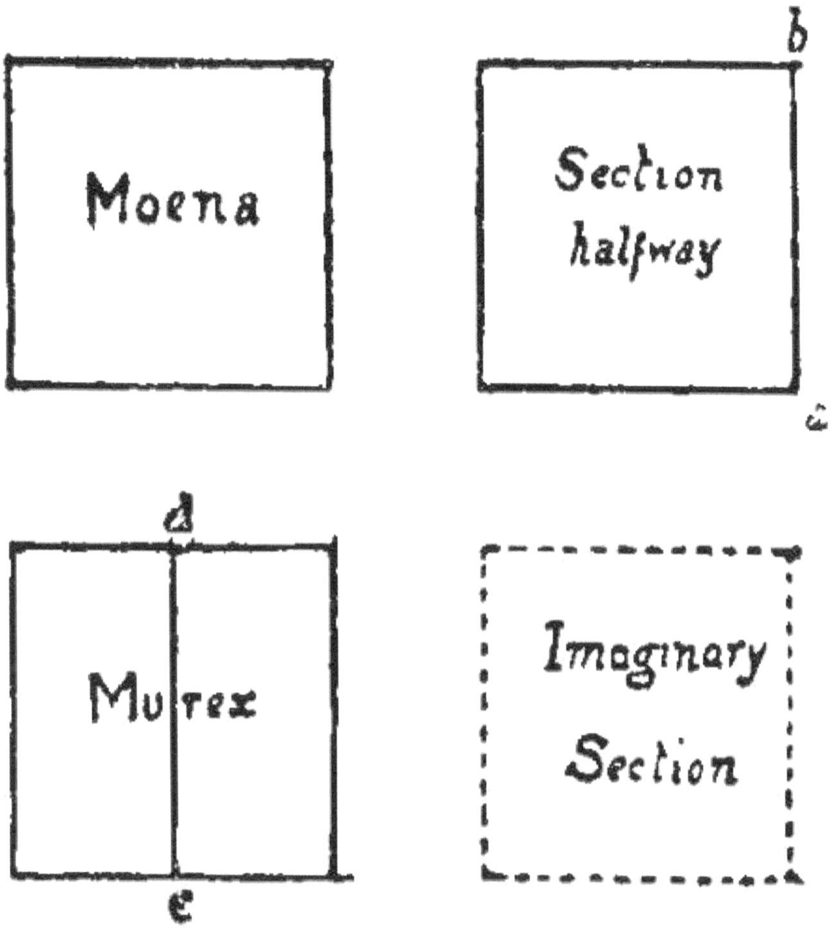


Fig. 22.

If he wanted the figure made by

Z
0

X
1⁄2

Y

1/2

it would be a little more difficult. He would have to take Moena, a section halfway between Moena and Murex, Murex and another square which he would have to regard as an *imaginary* section half-a-unit further Y than Murex (Fig. 22). He might now draw a ground plan of the sections; that is, he would draw Syce, and produce Cuspis and Dos half-a-unit beyond Nugæ and Cista. He would see that Cadus and Bolus would be cut half-way, so that in the half-way section he would have the point a (Fig. 23), and in Murex the point c. In the imaginary section he would have g; but this he might disregard, since the cube goes no further than Murex. From the points c and a there would be lines going Z, so that Iter and Semita would be cut half-way.

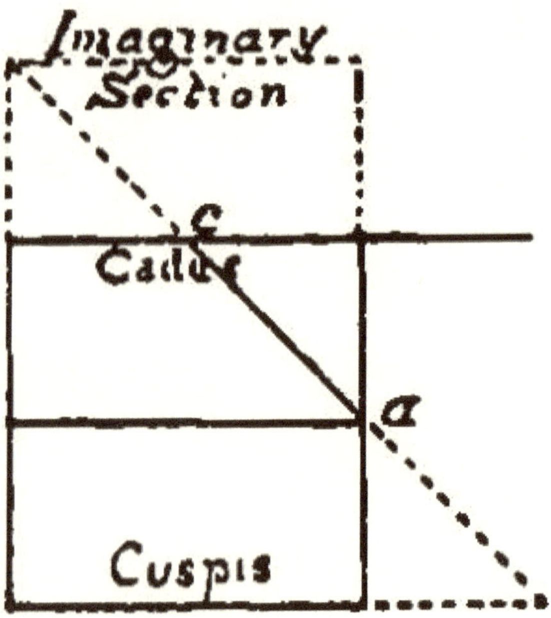


*Groundplan of Sections shown in Fig. 22.*

Fig. 23.

He could find out how far the two lines a b and c d (Fig. 22) are apart by referring d and b to Lama, and a and c to Crus.

In taking the third order of sections, a similar method may be followed.

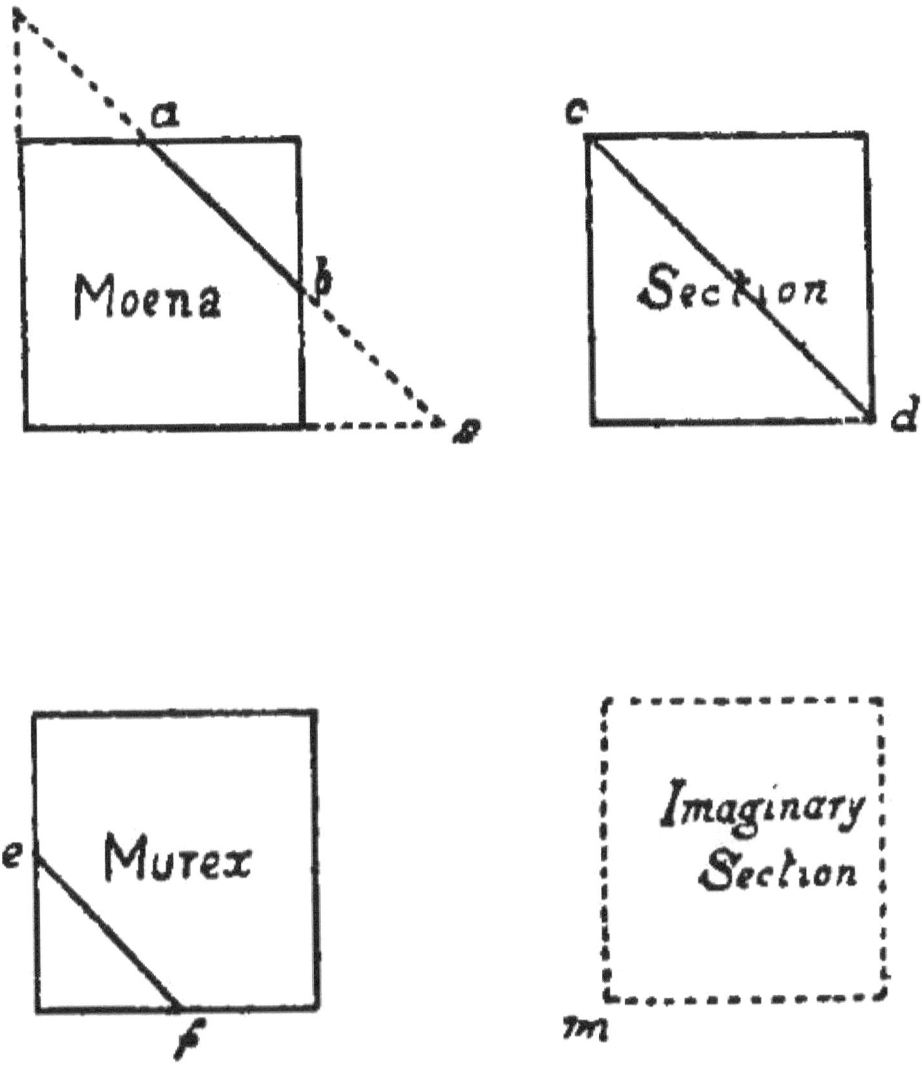


Fig. 24.

Suppose the sectional plane to cut Cuspis, Dos, and Arctos, each at one unit from Corvus. He would first take Moena, and as the sectional plane passes through Ilex and Nugæ, the line on Moena would be the diagonal passing through these two points. Then he would take Murex, and he would see that as the plane cuts Dos at one unit from Corvus, all he would have is the point Cista. So the whole figure is the Ilex to Nugæ diagonal, and the point Cista.

Now Cista and Ilex are each one inch from Corvus, and measured along lines at right angles to each other; therefore, they are d $(1)_2$ from each other. By referring Nugæ and Cista to Corvus he would find that they are also d $(1)_2$ apart; therefore the figure is an equilateral triangle, whose sides are each d $(1)_2$.

Suppose the sectional plane to pass through Mala, cutting Cuspis, Dos, and Arctos each at unit from Corvus. To find the figure, the plane-being would have to take Moena, a section half-way between Moena and Murex, Murex, and an imaginary section half-a-unit beyond Murex (Fig. 24). He would produce Arctos and Cuspis to points half-a-unit from Ilex and Nugæ, and by joining these points, he would see that the line passes through the middle points of Callis and Far (a, b, Fig. 24). In the last square, the imaginary section, there would be the point m; for this is 1½ unit from Corvus measured along Dos produced. There would also be lines in the other two squares, the section and Murex, and to find these he would have to make many observations. He found the points a and b (Fig. 24) by drawing a line from r to s, r and s being each 1½ unit from Corvus, and simply seeing that it cut Callis and Far at the middle point of each. He might now imagine a cube Mala turned about Arctos, so that Alvus came into his plane; he might then produce Arctos and Dos until they were each unit long, and join their extremities, when he would see that Via and Bucina are each cut half-way. Again, by turning Syce into his plane, and producing Dos and Cuspis to points 1½ unit from Corvus and joining the points, he would see that Bolus and Cadus are cut half-way. He has now determined six points on Mala, through which the plane passes, and by referring them in pairs to Ilex, Olus, Cista, Crus, Nugæ, Sors, he would find that each was d $(½)_2$ from the next; so he would know that the figure is an equilateral hexagon. The angles he would not have got in this observation, and they might be a serious difficulty to him. It should be observed that a similar difficulty does not come to us in our observation of the sections of a tessaract: for, if the angles of each side of a solid figure are determined, the solid angles are also determined.

There is another, and in some respects a better, way by which he might have found the sides of this figure. If he had noticed his plane-space much, he would have found out that, if a line be drawn to cut two other lines which meet, the ratio of the parts of the two lines cut off by the first line, on the side of the angle, is the same for those lines, and any other two that are parallel to them. Thus, if a b and a c (Fig. 25) meet, making an angle at a, and b c crosses them, and also crosses a' b' and a' c', these last two being parallel to a b and a c, then a b : a c :: a' b' : a' c'.

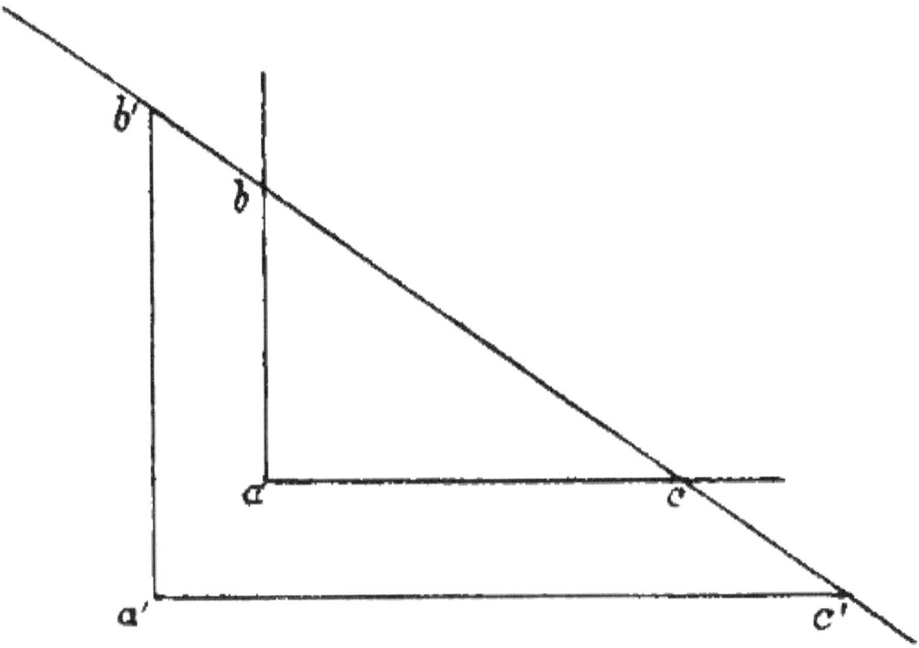


Fig. 25

If the plane-being knew this, he would rightly assume that if three lines meet, making a solid angle, and a plane passes through them, the ratio of the parts between the plane and the angle is the same for those three lines, and for any other three parallel to them.

In the case we are dealing with he knows that from Ilex to the point on Arctos produced where the plane cuts, it is half-a-unit; and as the Z, X, and Y lines are cut equally from Corvus, he would conclude that the X and Y lines are cut the same distance from Ilex as the Z line, that is half-a-unit. He knows that the X line is cut at 1½ units from Corvus; that is, half-a-unit from Nugæ: so he would conclude that the Z and Y lines are cut half-a-unit from Nugæ. He would also see that the Z and X lines from Cista are cut at half-a-unit. He has now six points on the cube, the middle points of Callis, Via, Bucina, Cadus, Bolus, and Far. Now, looking at his square sections, he would see on Moena a line going from middle of Far to middle of Callis, that is, a line d (½)2 long. On the section he would see a line from middle of Via to middle of Bolus d (1)2 long, and on Murex he would see a line from middle of Cadus to middle of Bucina, d (½)2 long. Of these three lines a b, c d, e f, (Fig. 24)—a b and e f are sides, and c d is a section of the required figure. He can find the distances between a and c by reference to Ilex, between b and d by reference to Nugæ, between c and e by reference to

Olus, and between d and f by reference to Crus; and he will find that these distances are each d (1⁄2)2.

Thus, he would know that the figure is an equilateral hexagon with its sides d (1⁄2)2 long, of which two of the opposite points (c and d) are d (1)2 apart, and the only figure fulfilling all these conditions is an equilateral and equiangular hexagon.

Enough has been said about sections of a cube, to show how a plane-being would find the shapes in any set as in

Z
I

.

X
II

.

Y
II

or

Z
I

.

X
I

.

Y
II

He would always have to bear in mind that the ratio of the lengths of the Z, X, and Y lines is the same from Corvus to the sectional plane as from any other point to the sectional plane. Thus, if he were taking a section where the plane cuts Arctos and Cuspis at one unit from Corvus and Dos at one-and-a-half, that is where the ratio of Z and of X to Y is as two to three, he would see that Dos itself is not cut at all; but from Cista to the point on Dos produced is half-a-unit; therefore from Cista, the Z and X lines will be cut at $\frac{2}{3}$ of $\frac{1}{2}$ unit from Cista.

It is impossible in writing to show how to make the various sections of a tessaract; and even if it were not so, it would be unadvisable; for the value of doing it is not in seeing the shapes themselves, so much as in the concentration of the mind on the tessaract involved in the process of finding them out.

Any one who wishes to make them should go carefully over the sections of a cube, not looking at them as he himself can see them, or determining them as he, with his three-dimensional conceptions, can; but he must limit his imagination to two dimensions, and work through the problems which a plane-being would have to work through, although to his higher mind they may be self-evident. Thus a three-dimensional being can see at a glance, that if a sectional plane passes through a cube at one unit each way from Corvus, the resulting figure is an equilateral triangle.

If he wished to prove it, he would show that the three bounding lines are the diagonals of equal squares. This is all a two-dimensional being would have to do; but it is not so evident to him that two of the lines are the diagonals of squares.

Moreover, when the figure is drawn, we can look at it from a point outside the plane of the figure, and can thus see it all at once; but he who has to look at it from a point in the plane can only see an edge at a time, or he might see two edges in perspective together.

Then there are certain suppositions he has to make. For instance, he knows that two points determine a line, and he assumes that three points determine a plane, although he cannot conceive any other plane than the one in which he exists. We assume that four points determine a solid space. Or rather, we say that *if* this supposition, together with certain others of a like nature, are true, we can find all the sections of a tessaract, and of other four-dimensional figures by an infinite solid.

When any difficulty arises in taking the sections of a tessaract, the surest way of overcoming it is to suppose a similar difficulty occurring to a two-dimensional being in taking the sections of a cube, and, step by step, to follow the solution he might obtain, and then to apply the same or similar principles to the case in point.

A few figures are given, which, if cut out and folded along the lines, will show some of the sections of a tessaract. But the reader is earnestly begged not to be content with *looking* at the shapes only. That will teach him nothing about a tessaract, or four-dimensional space, and will only tend to produce in his mind a feeling that "the fourth dimension" is an unknown and unthinkable region, in which any shapes may be right, as given sections of its figures, and of which any statement may be true. While, in fact, if it is the case that the laws of spaces of two and three dimensions may, with truth, be carried on into space of four dimensions; then the little our solidity (like the flatness of a plane-being) will allow us to learn of these shapes and relations, is no more a matter of doubt to us than what we learn of two- and three-dimensional shapes and relations.

There are given also sections of an octa-tessaract, and of a tetra-tessaract, the equivalents in four-space of an octahedron and tetrahedron.

A tetrahedron may be regarded as a cube with every alternate corner cut off. Thus, if Mala have the corner towards Corvus cut off as far as the points Ilex, Nugæ, Cista, and the corner towards Sors cut off as far as Ilex, Nugæ, Lama, and the corner towards Crus cut off as far as Lama, Nugæ, Cista, and the corner towards Olus cut off as far as Ilex, Lama, Cista, what is left of the cube is a tetrahedron, whose angles are at the points Ilex, Nugæ, Cista, Lama. In a similar manner, if every alternate corner of a tessaract be cut off, the figure that is left is a tetra-tessaract, which is a figure bounded by sixteen regular tetrahedrons.

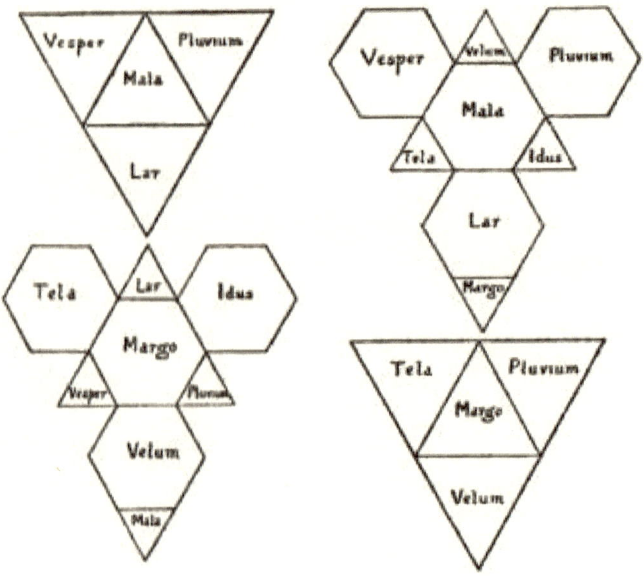


| Fig. 26. | Fig. 27. |
|---|---|
| Fig. 27. | Fig. 26. |

(i)

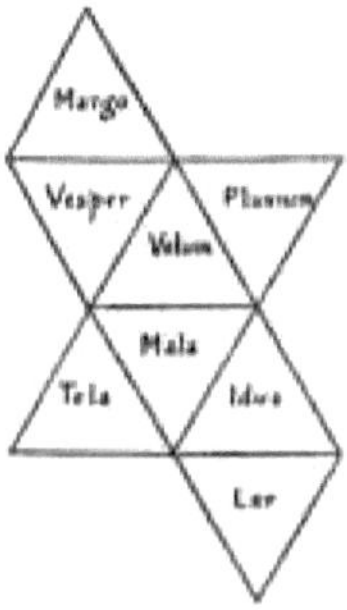

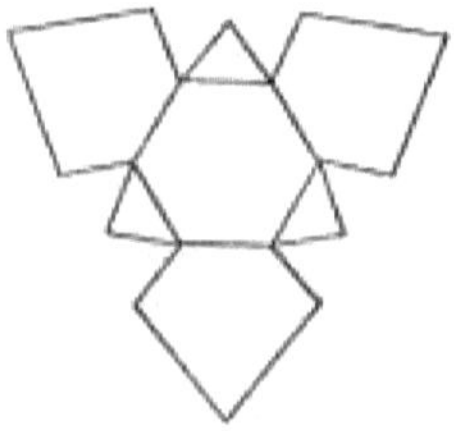

Fig. 28.

Fig. 29.

Fig. 30.

(ii)

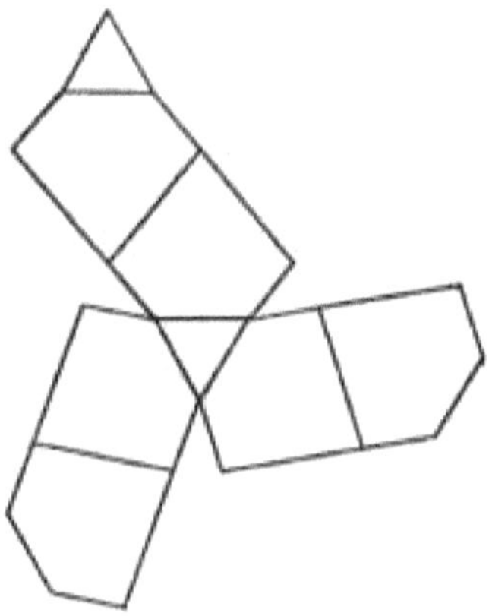

Fig. 31.

Fig. 32.

(iii)

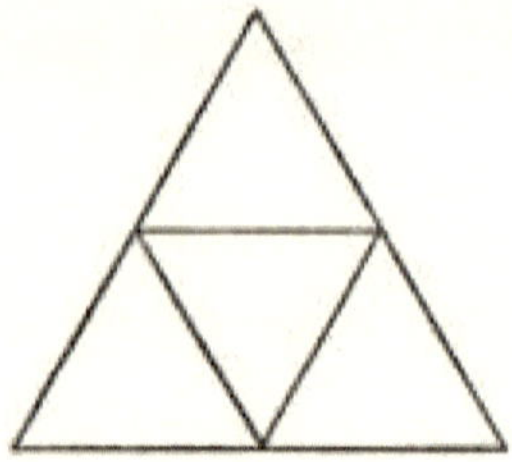

Fig. 33.

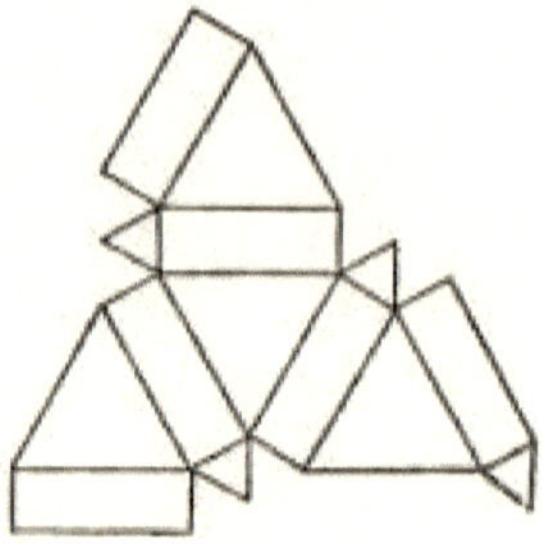

Fig. 34.

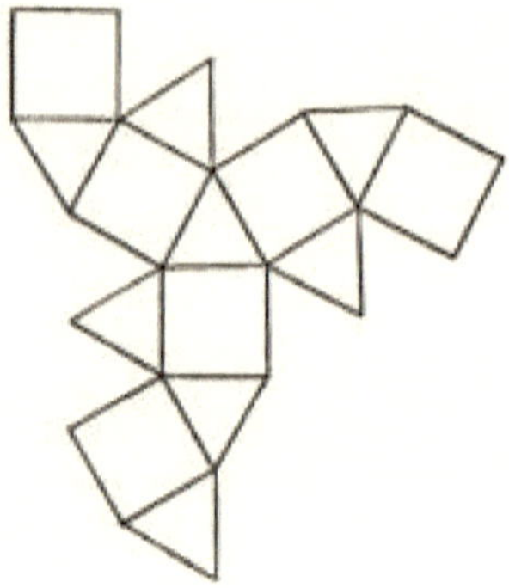

Fig. 35.

(iv)

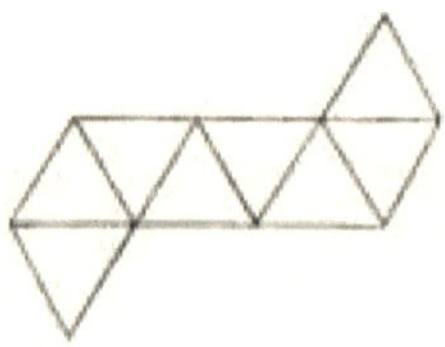

Fig. 36.

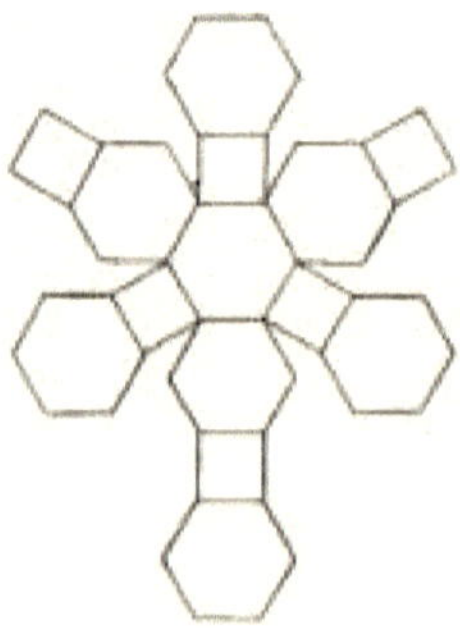

Fig. 37.

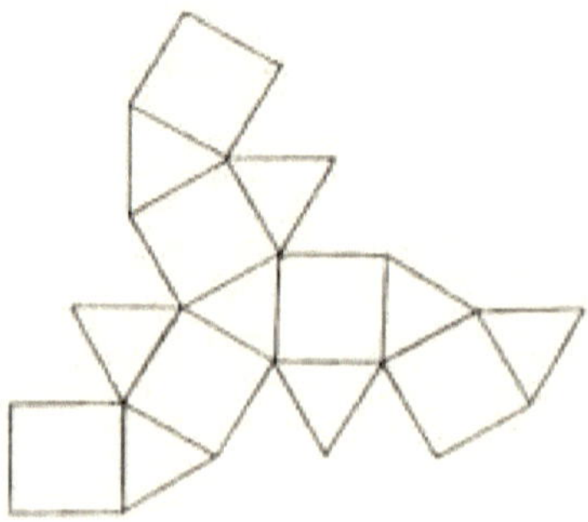

Fig. 38.

(v)

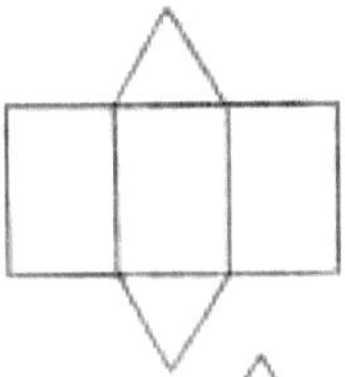

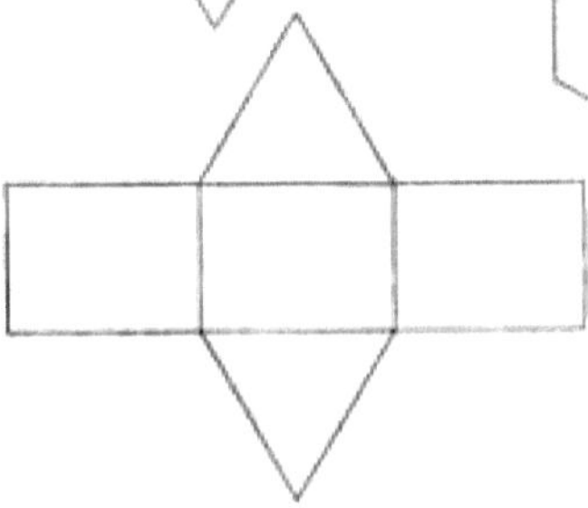

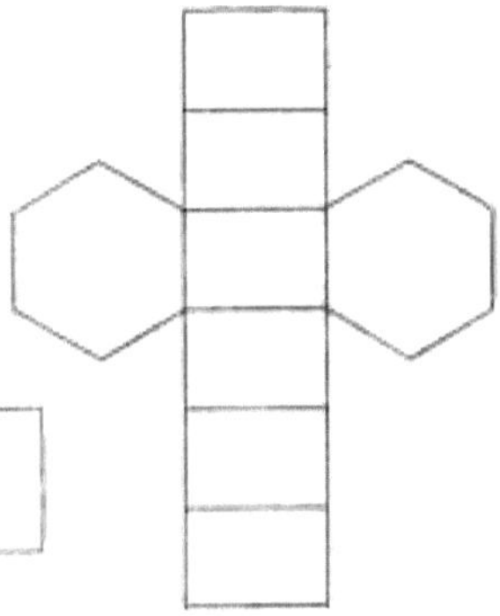

| Fig. 39. | Fig. 41. |
|---|---|
| Fig. 40. | |

(vi)

The octa-tessaract is got by cutting off every corner of the tessaract. If every corner of a cube is cut off, the figure left is an octa-hedron, whose angles are at the middle points of the sides. The angles of the octa-tessaract are at the middle points of its plane sides. A careful study of a tetra-hedron and an octa-hedron as they are cut out of a cube will be the best preparation for the study of these four-dimensional figures. It will be seen that there is much to learn of them, as—How many planes and lines there are in each, How many solid sides there are round a point in each.

## A DESCRIPTION OF FIGURES 26 TO 41.

| | | | | | Z | X | Y | W |
|---|---|---|---|---|---|---|---|---|
| Z X Y W | 26 | is a | section | taken | 1 . | 1 . | 1 . | 1 |
| - I . I . I . I | 27 | ” | ” | ” | 1√2 . | 1√2 . | 1√2 . | 1√2 |
| | 28 | ” | ” | ” | 2 . | 2 . | 2 . | 2 |
| | | | | | Z | X | Y | W |
| Z X Y W | 2 | i | sectio | take | 1 . | 1 . | 1 . | ½ |

9 s n n
. . . . I a
I I I 3
I I I 0 " " " 1√ . 1√ . 1√ . 3/4
2 2 2
3
1 " " " 2 . 2 . 2 . 1

3
2 " "" 2√ . 2√ . 2√ . 1√
2 2 2 4

The above are sections of a tessaract. Figures 33 to 35 are of a tetra-tessaract. The tetra-tessaract is supposed to be imbedded in a tessaract, and the sections are taken through it, cutting the Z, X and Y lines equally, and corresponding to the figures given of the sections of the tessaract.

Figures 36, 37, and 38 are similar sections of an octa-tessaract.

Figures 39, 40, and 41 are the following sections of a tessaract.

| Z X Y W | | | Z | X | Y | W |
|---|---|---|---|---|---|---|
| | 39 | is a section taken | 0 | . ½ | . ½ | . ½ |
| O I I I | 40 | " " " | 0 | . 1 | . 1 | . 1 |
| . . . . | 41 | " " " | 0 | . 1½ | . 1½ | . 1½ |

It is clear that there are four orders of sections of every four-dimensional figure; namely, those beginning with a solid, those beginning with a plane, those beginning with a line, and those beginning with a point. There should be little difficulty in finding them, if the sections of a cube with a tetra-hedron, or an octa-hedron enclosed in it, are carefully examined.

# APPENDIX K.

## MODEL 1. MALA.

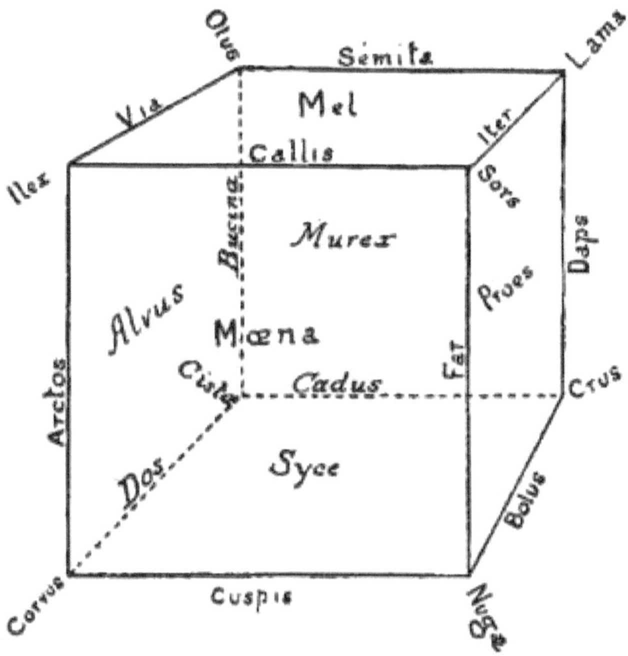


COLOURS: MALA, LIGHT-BUFF.

*Points*: Corvus, Gold. Nugæ, Fawn. Crus, Terra-cotta. Cista, Buff. Ilex, Light-blue. Sors, Dull-purple. Lama, Deep-blue. Olus, Red.

*Lines*: Cuspis, Orange. Bolus, Crimson. Cadus, Green-grey. Dos, Blue. Arctos, Brown. Far, French-grey. Daps, Dark-slate.

Bucina, Green. Callis, Reddish. Iter, Bright-blue. Semita, Leaden. Via, Deep-yellow.

*Surfaces*: Moena, Dark-blue. Proes, Blue-green. Murex, Light-yellow. Alvus, Vermilion. Mel, White. Syce, Black.

MODEL 2. MARGO.

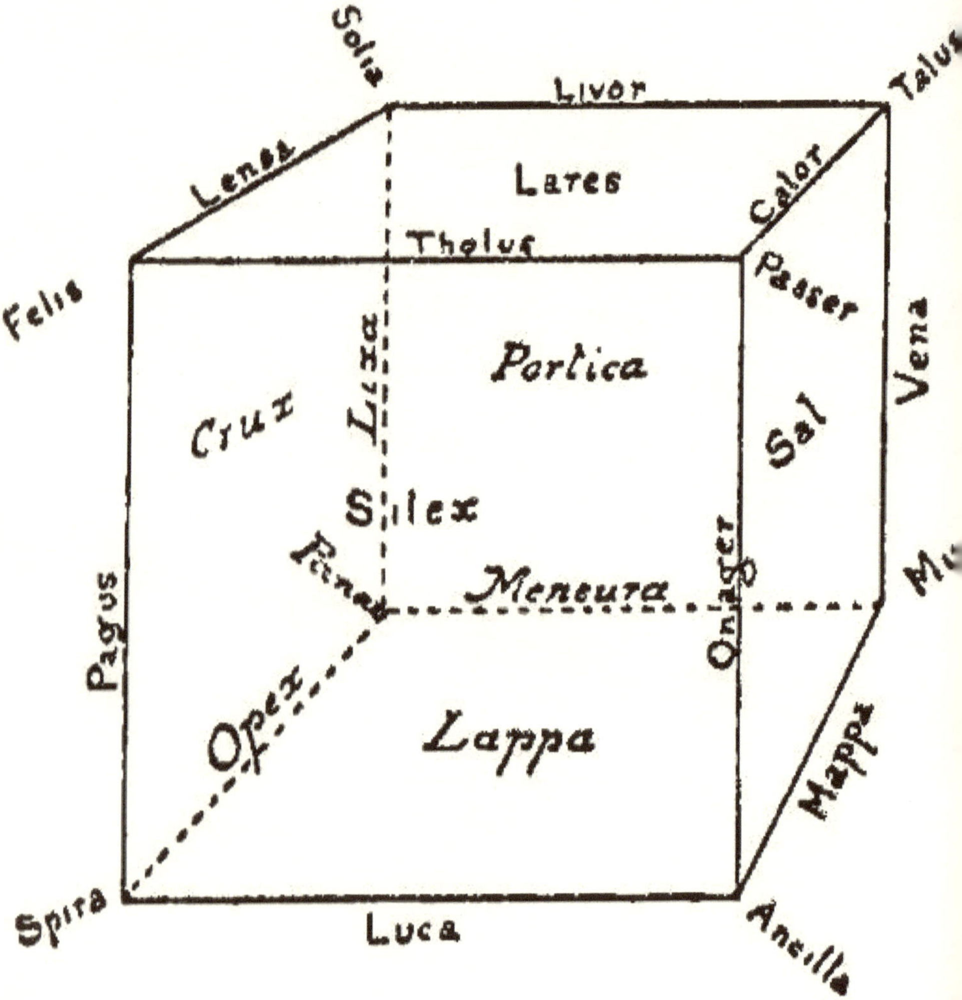


COLOURS: MARGO, SAGE-GREEN.

*Points*: Spira, Silver. Ancilla, Turquoise. Mugil, Earthen. Panax, Blue-tint. Felis, Quaker-green. Passer, Peacock-blue. Talus, Orange-vermilion. Solia, Purple.

*Lines*: Luca, Leaf-green. Mappa, Dull-green. Mensura, Dark-purple. Opex, Purple-brown. Pagus, Dull-blue. Onager, Dark-pink. Vena, Pale-pink. Lixa, Indigo. Tholus, Brown-green. Calor, Dark-green. Livor, Pale-yellow. Lensa, Dark.

*Surfaces*: Silex, Burnt-sienna. Sal, Yellow-ochre. Portica, Dun. Crux, Indian-red. Lares, Light-grey. Lappa, Bright-green.

## MODEL 3. LAR.

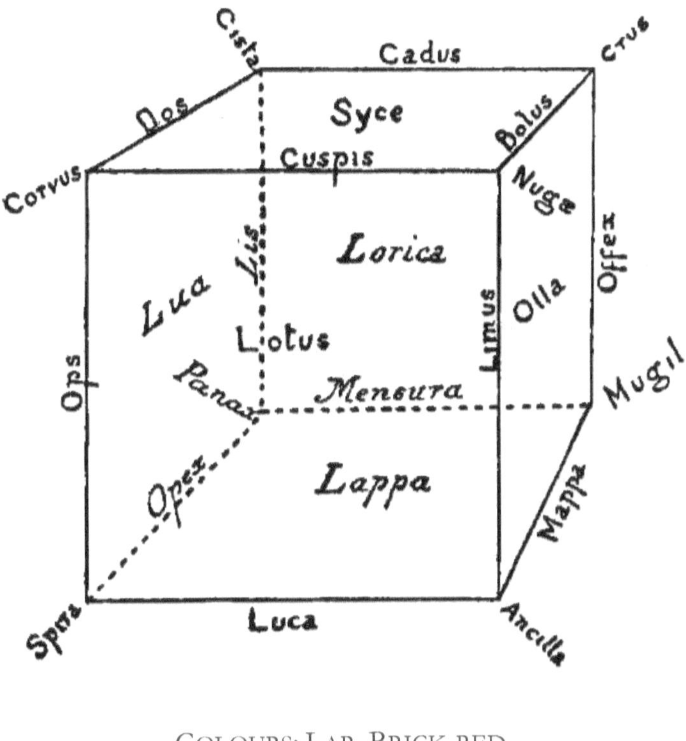


COLOURS: LAR, BRICK-RED.

*Points*: Spira, Silver. Ancilla, Turquoise. Mugil, Earthen. Panax, Blue-tint. Corvus, Gold. Nugæ, Fawn. Crus, Terra-cotta. Cista, Buff.

*Lines*: Luca, Leaf-green. Mappa, Dull-green. Mensura, Dark-purple. Opex, Purple-brown. Ops, Stone. Limus, Smoke. Offex,

Magenta. Lis, Light-green. Cuspis, Orange. Bolus, Crimson. Cadus, Green-grey. Dos, Blue.

*Surfaces*: Lotus, Azure. Olla, Rose. Lorica, Sea-blue. Lua, Bright-brown. Syce, Black. Lappa, Bright-green.

## MODEL 4. VELUM.

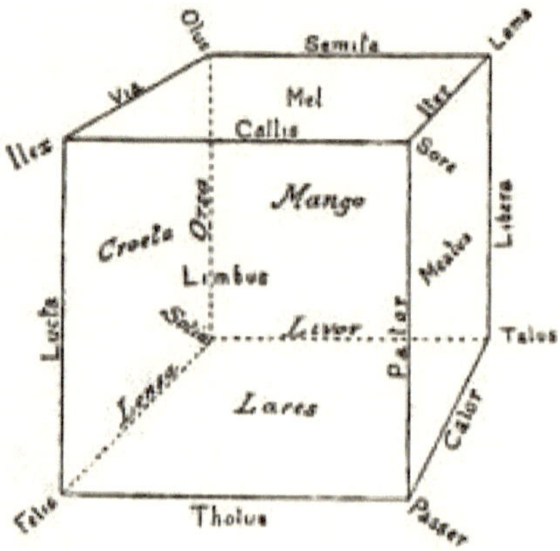


COLOURS: VELUM, CHOCOLATE.

*Points*: Felis, Quaker-green. Passer, Peacock-blue. Talus, Orange-vermilion. Solia, Purple. Ilex, Light-blue. Sors, Dull-purple. Lama, Deep-blue. Olus, Red.

*Lines*: Tholus, Brown-green. Calor, Dark-green. Livor, Pale-yellow. Lensa, Dark. Lucta, Rich-red. Pator, Green-blue. Libera, Sea-green. Orsa, Emerald. Callis, Reddish. Iter, Bright-blue. Semita, Leaden. Via, Deep-yellow.

*Surfaces*: Limbus, Ochre. Meatus, Deep-brown. Mango, Deep-green. Croeta, Light-red. Mel, White. Lares, Light-grey.

## MODEL 5. VESPER.

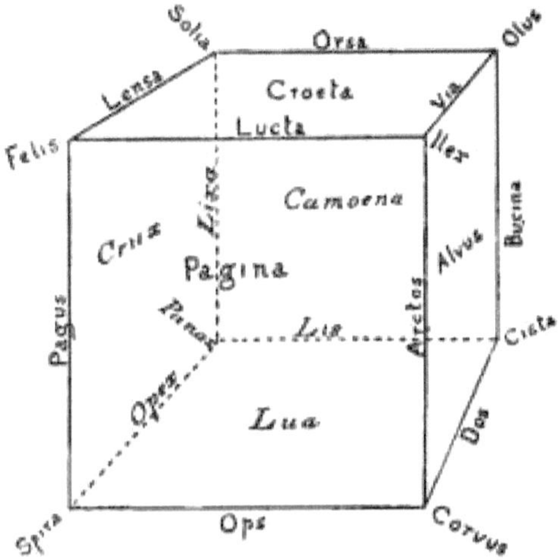


COLOURS: VESPER, PALE-GREEN.

*Points*: Spira, Silver. Corvus, Gold. Cista, Buff. Panax, Blue-tint. Felis, Quaker-green. Ilex, Light-blue. Olus, Red. Solia, Purple.

*Lines*: Ops, Stone. Dos, Blue. Lis, Light-green. Opex, Purple-brown. Pagus, Dull-blue. Arctos, Brown. Bucina, Green. Lixa, Indigo. Lucta, Rich-red. Via, Deep-yellow. Orsa, Emerald. Lensa, Dark.

*Surfaces*: Pagina, Yellow. Alvus, Vermilion. Camoena, Deep-crimson. Crux, Indian-red. Croeta, Light-red. Lua, Light-brown.

MODEL 6. IDUS.

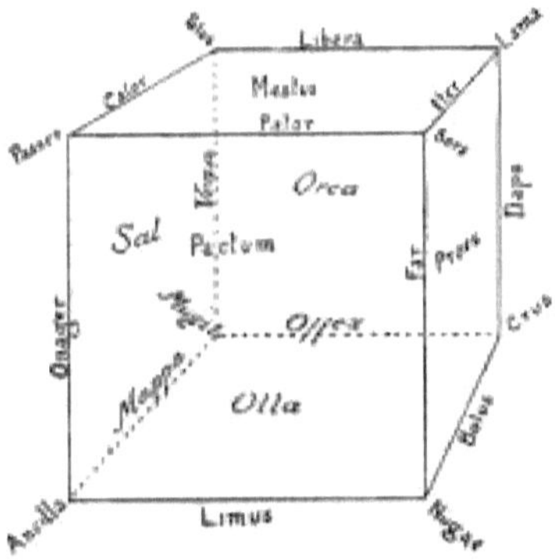


COLOURS: IDUS, OAK.

*Points*: Ancilla, Turquoise. Nugæ, Fawn. Crus, Terra-cotta. Mugil, Earthen. Passer, Peacock-blue. Sors, Dull-purple. Lama, Deep-blue. Talus, Orange-vermilion.

*Lines*: Limus, Smoke. Bolus, Crimson. Offex, Magenta. Mappa, Dull-green. Onager, Dark-pink. Far, French-grey. Daps, Dark-slate. Vena, Pale-pink. Pator, Green-blue. Iter, Bright-blue. Libera, Sea-green. Calor, Dark-green.

*Surfaces*: Pactum, Yellow-green. Proes, Blue-green. Orca, Dark-grey. Sal, Yellow-ochre. Meatus, Deep-brown. Olla, Rose.

## MODEL 7. PLUVIUM.

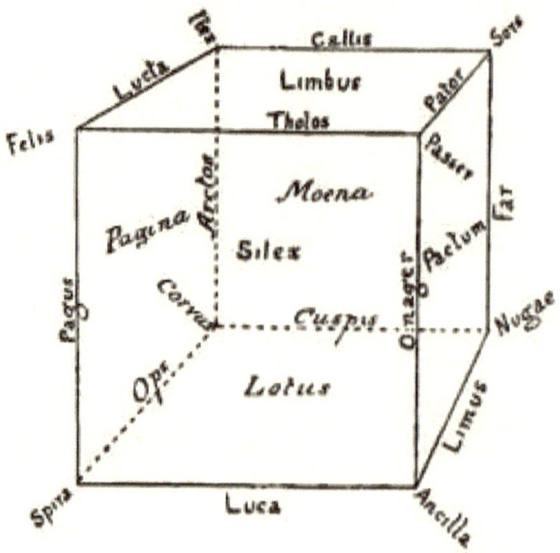


COLOURS: PLUVIUM, DARK-STONE.

*Points*: Spira, Silver. Ancilla, Turquoise. Nugæ, Fawn. Corvus, Gold. Felis, Quaker-green. Passer, Peacock-blue. Sors, Dull-purple. Ilex, Light-blue.

*Lines*: Luca, Leaf-green. Limus, Smoke. Cuspis, Orange. Ops, Stone. Pagus, Dull-blue. Onager, Dark-pink. Far, French-grey. Arctos, Brown. Tholos, Brown-green. Pator, Green-blue. Callis, Reddish. Lucta, Rich-red.

*Surfaces*: Silex, Burnt-Sienna. Pactum, Yellow-green. Moena, Dark-blue. Pagina, Yellow. Limbus, Ochre. Lotus, Azure.

## MODEL 8. TELA.

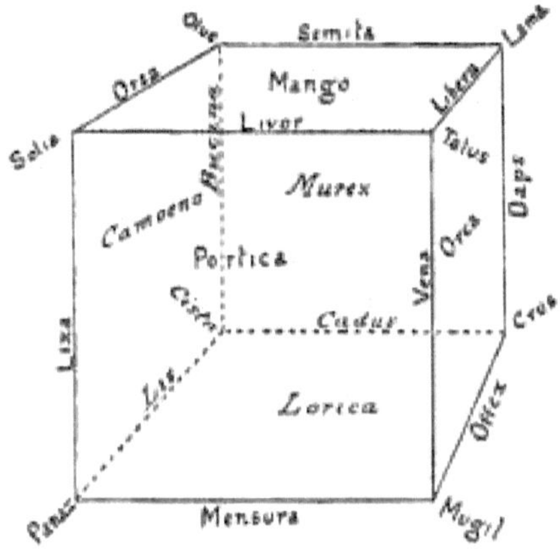


COLOURS: TELA, SALMON.

*Points*: Panax, Blue-tint. Mugil, Earthen. Crus, Terra-cotta. Cista, Buff. Solia, Purple. Talus, Orange-vermilion. Lama, Deep-blue. Olus, Red.

*Lines*: Mensura, Dark-purple. Offex, Magenta. Cadus, Green-grey. Lis, Light-green. Lixa, Indigo. Vena, Pale-pink. Daps, Dark-slate. Bucina, Green. Livor, Pale-yellow. Libera, Sea-green. Semita, Leaden. Orsa, Emerald.

*Surfaces*: Portica, Dun. Orca, Dark-grey. Murex, Light-yellow. Camoena, Deep-crimson. Mango, Deep-green. Lorica, Sea-blue.

## MODEL 9. SECTION BETWEEN MALA AND MARGO.

COLOURS: INTERIOR OR TESSARACT, WOOD.

*Points* (*Lines*): Ops, Stone. Limus, Smoke. Offex, Magenta. Lis, Light-green. Lucta, Rich-red. Pator, Green-blue. Libera, Sea-green. Orsa, Emerald.

*Lines* (*Surfaces*): Lotus, Azure. Olla, Rose. Lorica, Sea-blue. Lua Bright-brown. Pagina, Yellow. Pactum, Yellow-green. Orca, Dark-grey. Camoena, Deep-crimson. Limbus, Ochre. Meatus, Deep-brown. Mango, Deep-green. Croeta, Light red.

*Surfaces* (*Solids*): Pluvium, Dark-stone. Idus, Oak. Tela, Salmon. Vesper, Pale-green. Velum, Chocolate. Lar, Brick-red.

## MODEL 10. SECTION BETWEEN LAR AND VELUM.

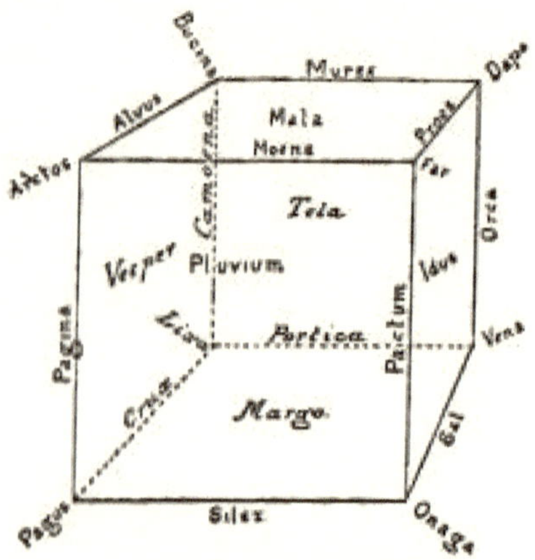


COLOURS: INTERIOR OR TESSARACT, WOOD.

*Points* (*Lines*): Pagus, Dull-blue. Onager, Dark-pink. Vena, Pale-pink. Lixa, Indigo. Arctos, Brown. Far, French-grey. Daps, Dark-slate. Bucina, Green.

*Lines* (*Surfaces*): Silex, Burnt-sienna. Sal, Yellow-ochre. Portica, Dun. Crux, Indian-red. Pagina, Yellow. Pactum, Yellow-green. Orca, Dark-grey. Camoena, Deep-crimson. Moena, Dark-blue. Proes, Blue-green. Murex, Light-yellow. Alvus, Vermilion.

*Surfaces* (*Solids*): Pluvium, Dark-stone. Idus, Oak. Tela, Salmon. Vesper, Pale-green. Mala, Light-buff. Margo, Sage-green.

## MODEL 11. SECTION BETWEEN VESPER AND IDUS.

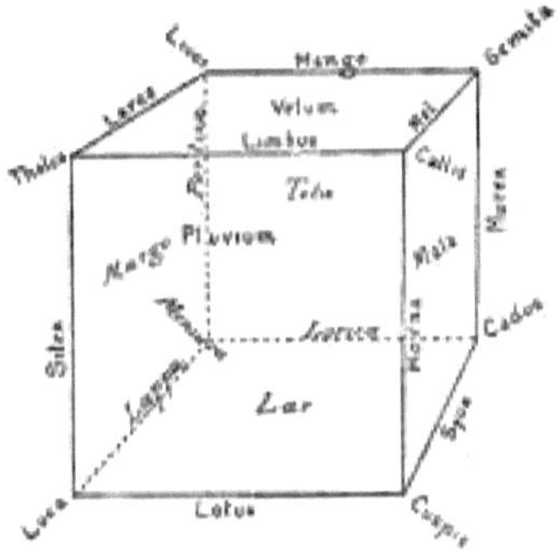


COLOURS: INTERIOR OR TESSARACT, WOOD.

*Points* (*Lines*): Luca, Leaf-green. Cuspis, Orange. Cadus, Green-grey. Mensura, Dark-purple. Tholus, Brown-green. Callis, Reddish. Semita, Leaden. Livor, Pale-yellow.

*Lines* (*Surfaces*): Lotus, Azure. Syce, Black. Lorica, Sea-blue. Lappa, Bright-green. Silex, Burnt-sienna. Moena, Dark-blue. Murex, Light-yellow. Portica, Dun. Limbus, Ochre. Mel, White. Mango, Deep-green. Lares, Light-grey.

*Surfaces* (*Solids*): Pluvium, Dark-stone. Mala, Light-buff. Tela, Salmon. Margo, Sage-green. Velum, Chocolate. Lar, Brick-red.

## MODEL 12. SECTION BETWEEN PLUVIUM AND TELA.

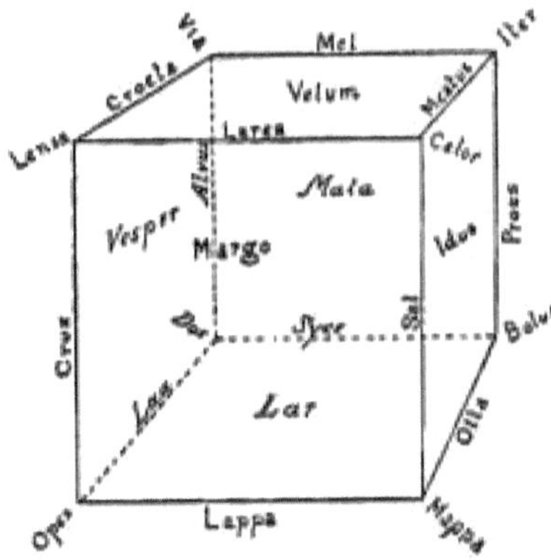


COLOURS: INTERIOR OR TESSARACT, WOOD.

*Points* (*Lines*): Opex, Purple-brown. Mappa, Dull-green. Bolus, Crimson. Dos, Blue. Lensa, Dark. Calor, Dark-green. Iter, Bright-blue. Via, Deep-yellow.

*Lines* (*Surfaces*): Lappa, Bright-green. Olla, Rose. Syce, Black. Lua, Bright-brown. Crux, Indian-red. Sal, Yellow-ochre. Proes, Blue-green. Alvus, Vermilion. Lares, Light-grey. Meatus, Deep-brown. Mel, White. Croeta, Light-red.

*Surfaces* (*Solids*): Margo, Sage-green. Idus, Oak. Mala, Light-buff. Vesper, Pale-green. Velum, Chocolate. Lar, Brick-red.

www.ingramcontent.com/pod-product-compliance
Lightning Source LLC
Chambersburg PA
CBHW031325230926
51848CB00001B/93
*9789373499024*